Jim Kallarackal

The Higgs boson resonance

Jim Kallarackal

The Higgs boson resonance
A chirally invariant Higgs-Yukawa model in lattice field theory

Südwestdeutscher Verlag für Hochschulschriften

Impressum/Imprint (nur für Deutschland/only for Germany)
Bibliografische Information der Deutschen Nationalbibliothek: Die Deutsche Nationalbibliothek verzeichnet diese Publikation in der Deutschen Nationalbibliografie; detaillierte bibliografische Daten sind im Internet über http://dnb.d-nb.de abrufbar.
Alle in diesem Buch genannten Marken und Produktnamen unterliegen warenzeichen-, marken- oder patentrechtlichem Schutz bzw. sind Warenzeichen oder eingetragene Warenzeichen der jeweiligen Inhaber. Die Wiedergabe von Marken, Produktnamen, Gebrauchsnamen, Handelsnamen, Warenbezeichnungen u.s.w. in diesem Werk berechtigt auch ohne besondere Kennzeichnung nicht zu der Annahme, dass solche Namen im Sinne der Warenzeichen- und Markenschutzgesetzgebung als frei zu betrachten wären und daher von jedermann benutzt werden dürften.

Verlag: Südwestdeutscher Verlag für Hochschulschriften GmbH & Co. KG
Heinrich-Böcking-Str. 6-8, 66121 Saarbrücken, Deutschland
Telefon +49 681 37 20 271-1, Telefax +49 681 37 20 271-0
Email: info@svh-verlag.de

Approved by: Berlin, Humboldt Universität zu Berlin, Diss., 2011

Herstellung in Deutschland:
Schaltungsdienst Lange o.H.G., Berlin
Books on Demand GmbH, Norderstedt
Reha GmbH, Saarbrücken
Amazon Distribution GmbH, Leipzig
ISBN: 978-3-8381-3094-1

Imprint (only for USA, GB)
Bibliographic information published by the Deutsche Nationalbibliothek: The Deutsche Nationalbibliothek lists this publication in the Deutsche Nationalbibliografie; detailed bibliographic data are available in the Internet at http://dnb.d-nb.de.
Any brand names and product names mentioned in this book are subject to trademark, brand or patent protection and are trademarks or registered trademarks of their respective holders. The use of brand names, product names, common names, trade names, product descriptions etc. even without a particular marking in this works is in no way to be construed to mean that such names may be regarded as unrestricted in respect of trademark and brand protection legislation and could thus be used by anyone.

Publisher: Südwestdeutscher Verlag für Hochschulschriften GmbH & Co. KG
Heinrich-Böcking-Str. 6-8, 66121 Saarbrücken, Germany
Phone +49 681 37 20 271-1, Fax +49 681 37 20 271-0
Email: info@svh-verlag.de

Printed in the U.S.A.
Printed in the U.K. by (see last page)
ISBN: 978-3-8381-3094-1

Copyright © 2012 by the author and Südwestdeutscher Verlag für Hochschulschriften GmbH & Co. KG and licensors
All rights reserved. Saarbrücken 2012

Meiner Liebe, Martina,
und meinen Eltern gewidmet

Contents

1	**Introduction**	**1**
2	**Definition of a chirally invariant Higgs-Yukawa model**	**7**
	2.1 The continuum formulation and its symmetries	9
	2.2 The model on a discretized space-time lattice	14
	2.3 Simulation strategy	22
3	**Analytic properties and perturbative calculations**	**39**
	3.1 Perturbative expansion in the continuum	39
	3.1.1 The Higgs boson propagator	42
	3.1.2 The Goldstone boson propagator	51
	3.2 Lattice perturbation theory	54
4	**Resonance parameters of the Higgs boson**	**61**
	4.1 Mass bounds of the Higgs boson	63
	4.1.1 Observables and Higgs boson mass bounds	64
	4.2 Resonance mass and width of the Higgs boson	67
	4.2.1 The scattering phase in the continuum and in a finite box	68
	4.2.2 Numerical results	78
5	**Beyond the Standard model: A fourth generation of fermions**	**89**
	5.1 The model	91
	5.2 Current Mass bounds related with a fourth generation	93
	5.3 Observables and extraction of mass eigenvalues	94
	5.3.1 Extraction of mass eigenvalues	96

5.4	Cut off dependent Higgs boson mass bounds	99
	5.4.1 Numerical results	100
5.5	Higgs boson mass bounds with varying top quark masses	103
	5.5.1 Numerical results	104
5.6	Conclusion	105

6 Summary and conclusion   **109**

Appendix A: Lattice Parametrization   **117**

Appendix B: Perturbative calculations   **121**

Appendix C: Two Particle Energy Levels   **133**

Acknowledgments   **139**

Bibliography   **144**

Own Publications   **146**

List of Figures   **148**

List of Tables   **149**

1 Introduction

The emergence of quantum mechanics and the theory of special relativity opened the door to a modern perspective of physics leaving the classical Newtonian physics behind. Both theories required a new interpretation of physical quantities. The special theory of relativity embeds space and time in a four dimensional space-time. The fundamental principles of relativity are reflected in a symmetry, namely the Poincaré invariance of the action. The early days of quantum mechanics suffered from the lack of a consistent interpretation of particles and their paths. The understanding of quantum mechanics was a long journey of mistakes and illumination; the initial path is reflected in [38].

The unification of the special theory of relativity and quantum mechanics inevitably led to relativistic quantum field theory. Within the classical theory, the electromagnetic forces were already formulated in a way that was easy to incorporate with special relativity. Eventually it was Dirac who accomplished to provide a linear differential equation, which allowed to describe electrons within a quantum field theory. The symmetries of the action, which describes the relevant physics and helped to formulate the principles of relativity, turned into a more general idea and strongly constrained the possible interactions of elementary particles. Local symmetries, also known as local gauge invariance, enabled to incorporate all underlying symmetries of the original classical theory. Furthermore, the electromagnetic potential emerged as an additional degree of freedom. These degrees of freedom within a relativistic quantum field theory are known as gauge bosons. In the case of quantum electromagnetism, the gauge boson is the photon and it belongs to the simplest symmetry group within the standard model of particle physics, the $U_{em}(1)$.

The relativistic quantum theory of electrodynamics also revealed another aspect, namely the need for renormalization. While the theory was able to predict experimental results at leading order of perturbation theory in the electromagnetic coupling, higher orders involved infinities and could not generate meaningful results. The distinction between the parameters of the theory and physical observables then led to the general theory of renormalization. In the modern view of quantum field theory, the symmetries of the model together with the requirement of renormalizability restrict the structure of mutual interactions in such a way that the most general form of such a theory coincides with the empirically motivated model.

Improvements in experimental measurements and observations of atomic interactions allowed to arrange all known interactions within four fundamental forces. The electromagnetic, weak, strong and gravitational force. While the gravitational force acts on large distances, the others are microscopic and describe interactions of electrons and the constituents of protons and neutrons. The standard model of particle physics embeds the first three forces in a relativistic quantum field theory and its predictions and implications have been extensively tested in the last decades. The strong interaction describes quarks and gluons, which are the constituents of protons and so called mesons. However, it turns out that the quarks are not exposed in the observable spectrum of the theory but that they form bound states and excited states thereof which constitute the observed heavy particles, the so called baryons and mesons.

The symmetry of the standard model is given by the group structure $SU_W(2) \times U_Y(1) \times SU(3)$. The elementary particles and their transformation property under the above symmetry together with the requirement of renormalizability then define the standard model of particle physics. An overview of the particle content is presented in figure 1.1. The leptons and quarks build up the ordinary matter. The leptons are arranged within a $SU_W(2)$ doublet and their mutual interaction is dictated by the weak $SU_W(2) \times U_Y(1)$ symmetry. The gauge bosons of the weak symmetry are the massive $W^\pm$ and the Z boson. The quark fields are also arranged within a $SU_W(2)$ doublet and thus they interact with the weak gauge bosons. Moreover, each quark is arranged within a $SU(3)$ vector. The interactions induced by the $SU(3)$ symmetry describe the strong interaction which is mediated by the gluons. The strong interaction confines the quarks in bound states. These bound states finally build up the known spectrum of hadronic particles such as the proton, the neutron, or the π mesons. The Higgs particle in figure 1.1 is highlighted and indicates that of all particles embedded within the standard model of particle physics, the Higgs boson is the only one which has not yet been observed. The Higgs boson plays a crucial role in the generation of masses for the fermions. It is known that all leptons, quarks and weak gauge bosons are massive.

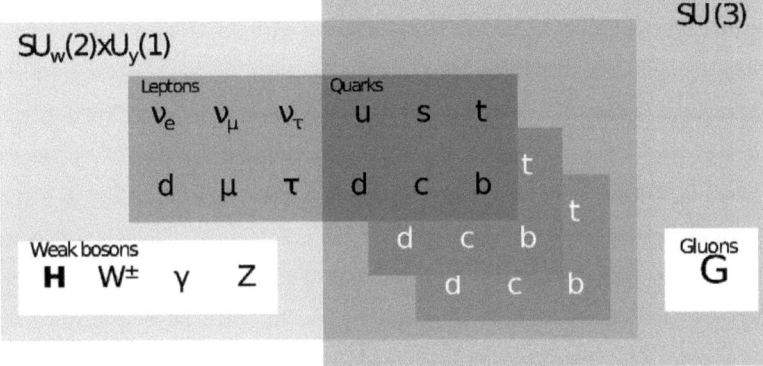

Figure 1.1: The figure shows the particle content of the standard model of elementary particle physics. The leptons and quarks correspond to the usual matter. While quarks are confined in bound states such as the proton or the neutron, leptons can be observed directly in high energy experiments. The photon, the weak gauge bosons and the gluons mediate the electromagnetic, the weak and the strong nuclear forces between the fermions. The Higgs boson, highlighted in red, plays a crucial role in the generation of masses for the fundamental particles. It has not been observed yet; its mass and resonance width are the main topic of this thesis.

Despite its outstanding success, a major key in the framework of the standard model is yet lacking to be confirmed experimentally. The symmetries of the electroweak sector of the theory do not permit massive fermions or massive gauge bosons. It is experimentally known that the weak gauge bosons as well as the leptons and the quarks are massive. The weak gauge bosons mediate the weak force similar to the photon mediating the electromagnetic force. As a consequence, the symmetry in the weak sector of the standard model must be broken. The above fact is incorporated into the model by the concept of spontaneous symmetry break down, which is modelled with the scalar sector by involving the Higgs boson.

This work is dedicated to investigate the properties of the Higgs boson from first principles. In particular the Higgs boson mass bounds and its resonance width will be computed within the framework of lattice field theory. The path integral formulation of quantum field theory allows to access observables of the theory by numerical Monte Carlo integration techniques.

The model is restricted to the electroweak sector with its $SU_W(2) \times U_Y(1)$ symmetry. The largest couplings to the Higgs boson are the so called Yukawa couplings, which define the interactions between the fermions and the Higgs boson. Among all Yukawa couplings, it is the top quark whose coupling to the Higgs is by far larger (a factor of roughly 40) than all remaining Yukawa couplings. Furthermore, it is known that the Higgs boson couples weakly to the mediators of the weak forces (the $W^\pm, Z$ bosons). Though neglecting them alters the theory conceptually, it is expected that their influence on the mass of the Higgs boson is negligible. Scattering processes however, can be described by using the Goldstone bosons instead. The last statement is phrased in the Goldstone equivalence theorem and is made more clearer in Chapter 2. From the arguments above, it is expected that the restrictions may not have a significant influence on the final results. Since the investigation of chiral gauge theories on the lattice is still an open and demanding subject on its own, being able to neglect the gauge boson degrees of freedom is a great simplification.

Recently, an extensive study of the upper and lower Higgs boson mass bound was performed within this model [31, 30]. For the upper Higgs boson it was necessary to consider an analytic propagator, and the Higgs boson mass was extracted from the pole of the real part of the propagator. Indeed, the time dependence of the Higgs boson propagator which yields the time correlator is not sufficient in order to extract the Higgs boson mass. The Higgs boson is an elementary particle but it does not appear as an asymptotic state as it decays into any even number of Goldstone bosons respectively weak gauge bosons. The time slice correlator is then dominated by the masses of the Goldstone bosons and the Higgs boson mass lies within a continuous spectrum of two particle energies. Though the analysis of the propagator can determine the mass of the Higgs boson as a resonance by means of analytic continuation, it depends on an analytic function. A generic form of the Higgs boson propagator is not known and one has to utilize the functional form of the propagator suggested by one loop perturbation theory. A priori, one cannot know, whether at large couplings, other functions than in the one loop approximation may dominate the behaviour of the propagator. Furthermore, the resonance width, which is connected with the imaginary part of the complex Higgs boson propagator, is neglected. Hence, another method to extract the resonance mass is highly appreciated. Such a method is the finite size technique proposed in [46]. It is genuinely non perturbative and does not need any knowledge of the propagator. In three distinct physical setups both techniques will be contrasted in Chapter 4.

The second main focus of this work is to investigate the effect of a heavy fourth generation of quarks on the aforementioned Higgs boson mass bounds. An extension of the standard model with a fourth generation of heavy quarks and leptons, arranged within a $SU_W(2)$ doublet, permits to alter the model in a way such that it is compatible with electroweak precision measurements. A fourth generation of fermions provides various prospects to augment the model to enable a deeper understanding of flavor physics and mass hierarchies [40]. The main motivation, however, is that it may satisfy the three Sakharov conditions [55] such that the observed baryon asymmetry of the universe is consistent with theory [16, 41]. It is important to mention that within the standard model the Sakharov conditions cannot be realized when experimental constraints on the Higgs boson mass and CP violating phase in the CKM matrix are taken into account. Large Yukawa couplings give rise to potential non perturbative effects and the method at hand is perfectly suited to study such effects.

In the following, the structure of the chapters are summarized briefly. Chapter 2 introduces the model in continuous space time and the translation to a finite discrete space time lattice is discussed. Once the model is defined, its symmetries and the transformation properties of the fields in continuum and in finite volume are discussed in detail. Thereafter the simulation strategy is explained and the basic technique to extract mass eigenvalues is presented. The Källen-Lehman representation of the interacting two point function will play a role throughout this work and thus some details are given at the end of the chapter.

Chapter 3 gives explicit results on the one loop approximation of the scalar propagators. The result is used to fit numerical data and to extract the resonance mass. The chapter closes with perturbation theory on a finite discretized space time lattice. The calculations show an agreement of physical quantities obtained by means of the Monte Carlo simulation with those computed within bare perturbation theory. The one loop contribution of the Neuberger overlap operator is also computed and compared with numerical data to high precision for the first time. At small values of the Yukawa couplings, the perturbative result may be used to study the effect of the technical parameters of the overlap operator.

As announced before, Chapter 4 describes the finite volume technique in order to extract resonance parameters. In [54] a modification of the method was proposed. Both methods allow to access the scattering phases from Monte Carlo simulations. The aforementioned modification turned out to be very helpful as it allows to collect the necessary number of scattering phases in order to perform a good fit to extract the resonance parameters. Finally, the results obtained from the finite size techniques are compared with the results obtained from the analysis of the propagator.

Chapter 5 addresses the question of a fourth generation of heavy quarks. The Higgs boson mass bounds are explored in the presence of a heavy quark of roughly 700 GeV. The bounds are compared to those established for the standard model quarks.

2 Definition of a chirally invariant Higgs-Yukawa model

Within the electroweak standard model, the pure Higgs-Yukawa sector describes the interaction between fermions and scalar particles. This electroweak sector of the standard model plays a crucial role in the understanding of mass generation for fermion and W and Z bosons. In the complete electroweak standard model a local $SU(2)$ gauge symmetry is established in order to describe the weak interactions. Phenomenologically it is known that weak interactions treat left handed components of the fermions different than the right handed components. An explicit fermion mass term is therefore not allowed. Furthermore, it is known that the weak gauge bosons are massive, but mass terms for gauge bosons are not compatible with local gauge symmetry. The electroweak sector of the standard model utilizes the scalar sector in order to keep the $SU_W(2) \times U_Y(1)$ model manifestly invariant and renormalizable while at the same time allowing a mechanism, which provides a framework in which the weak gauge bosons as well as the fermions acquire a mass. One of the fundamental ingredients of this mass generation, the so called Higgs mechanism, is the necessity of the Higgs boson, which is supposed to appear as a resonance in the particle spectrum. Though theoretical predictions based on the electroweak theory have been verified in experiments in the last decades, the Higgs boson itself has not yet been observed. In addition, despite experimental and perturbation theory based constraints on the Higgs boson mass and width are available, much less is known about eventual non-perturbative properties concerning the Higgs boson resonance. Constraining the Higgs boson resonance parameters reliably also with non-perturbative calculations is therefore of great importance for phenomenology and for experiments, in particular in light of the just started LHC.

The main focus in this work is to explore potential non-perturbative features within the pure Higgs-Yukawa sector. The Higgs-Yukawa model considered here neglects all gauge boson interactions and considers only a degenerate fermion doublet. As explained below, both restrictions are reasonable with respect to the quantities which are of interest in this work. Furthermore, a chiral gauge theory involves conceptual difficulties and is beyond the scope of this work. As it is the aim to compute mass bounds and resonance parameters of the Higgs boson at a number of couplings, including possibly non-perturbative accessible values, the main contributions arise from the strongest couplings in the fermion-scalar sector. On the contrary, the weak coupling

$$\frac{g^2}{8M_W^2} = \frac{G_F}{\sqrt{2}} \Rightarrow g = 0.357$$
$$\tan\theta_W = \frac{g'}{g} \Rightarrow g' = 0.652$$

is known to be small and thus the contribution of gauge bosons to the mass of the Higgs boson is sub-dominant with respect to those stemming from the heavy top- and bottom-quark doublet as well as from large quartic couplings of the scalar self-interaction. Furthermore, the Goldstone equivalence theorem [53] states that the contributions of the $W^\pm$ and Z bosons to scattering amplitudes are identical to those of the Goldstone bosons in a theory without gauge bosons. This ensures to extract scattering phases reliably which in turn will be used to compute the resonance width and resonance mass and associate them to the resonance parameters of the Higgs boson in the electroweak sector of the standard model.

The standard model top quark is roughly a factor 40 larger than the mass of the bottom quark ($m_t = 171.2(2.1)$ GeV , $m_b = 4.20(0.17)$ GeV [8]). Of course, it would be physically more realistic to perform calculations with such a mass-splitting realized. However, while in the mass degenerate case ($m_t = m_b$) it can be ensured that the determinant of the fermion matrix when considered on a lattice is strictly real valued, a mass splitting within the fermion doublet allows for a complex phase in the fermion determinant which in turn is numerically difficult to compute. Nevertheless, the effects of a mass splitting, adjusted to the physical situation, on the lower Higgs boson mass bound has been investigated in [25]. All results presented in this work will however be based on a mass degenerate fermion doublet. Chapter 5 explores the mass bounds of the Higgs boson in the presence of a heavy mass degenerate fourth generation of quark doublets. What is of particular interest concerning the fourth generation of fermions is the *relative change* of the Higgs boson mass bounds. Since for a top quark mass of $m_t = 171.2(2.1)$ GeV also a mass degenerate quark doublet was considered, it suffices to use a mass degenerate doublet for the fourth generation quarks in order to quantify the relative shifts of the Higgs boson mass bounds.

In this chapter, first the continuum model is introduced and its symmetries are discussed. The translation of these continuum symmetries to a finite discretized space time lattice plays an important role in the investigation of Higgs-Yukawa models and thus some details on the transformation properties of the fields are discussed. Afterwards, the formulation of the Higgs-Yukawa model on the lattice is given and once again the symmetries are exposed. The symmetry on the lattice has to be modified but finally it yields the correct symmetries when the limit to zero lattice spacing is performed. The last part focuses on the simulation strategy. The line of constant physics as well as the method to determine the upper and the lower Higgs boson mass bound is explained. Finally, the observables which are used to compute the Higgs boson mass and the fermion mass are defined and some analytic details on the propagator is given in order to discuss the unstable nature of the Higgs boson for large quartic or large Yukawa couplings.

2.1 The continuum formulation and its symmetries

The Higgs-Yukawa model is defined by the Lagrangian and the corresponding generating functional for the Green functions of the theory. With regard to the later lattice version of the model, the Euclidean version of the model will be considered here. The particle content contains the scalar sector and the heaviest quark doublet consisting of the top and the bottom quark. Due to the neglect of gauge bosons the Lagrangian exhibits a *global* $SU_W(2) \times U_Y(1)$ inner symmetry rather than a local symmetry. The Euclidean action is given by

$$L_E^{HY} = \frac{1}{2}(\partial_\mu \varphi)^\dagger \cdot (\partial^\mu \varphi) + \frac{1}{2}m^2 \varphi^\dagger \cdot \varphi + \lambda\left(\varphi^\dagger \cdot \varphi\right)^2$$
$$+ \bar{t}\,\slashed{D}\,t + \bar{b}\,\slashed{D}\,b + y_b \begin{pmatrix} \bar{t} \\ \bar{b} \end{pmatrix}_L^T \cdot \varphi\, b_R + y_t \begin{pmatrix} \bar{t} \\ \bar{b} \end{pmatrix}_L^T \cdot \tilde{\varphi}\, t_R \quad + h.c.. \quad (2.1)$$

The scalar fields are defined on $\mathbb{R}^4$

$$\varphi : \mathbb{R}^4 \to \mathbb{C}$$

while the fermion fields are complex Grassmann fields. $\tilde{\varphi}$ transforms like a $SU_W(2)$ vector and is given by

$$\tilde{\varphi} := i\sigma_2 \varphi^* = \begin{pmatrix} \varphi_2^* \\ -\varphi_1^* \end{pmatrix}$$

$\slashed{D}$ is a shorthand notation for the contraction of the free Dirac operator with the gamma matrices

$$\slashed{D} = \gamma_\mu^E \partial_\mu$$

where γ_μ^E are the Euclidean gamma matrices which are explicitly given by

$$\gamma_{1,2,3} := -i\gamma_{1,2,3}^{\text{Minkowski}}$$
$$\gamma_4 := \gamma_0^{\text{Minkowski}}$$
$$\gamma_{1,2,3} = \begin{pmatrix} 0 & -i\sigma_{1,2,3} \\ i\sigma_{1,2,3} & 0 \end{pmatrix}$$

The Euclidean gamma matrices are hermitian $\gamma_\mu^\dagger = \gamma_\mu$. The Pauli matrices $\sigma_{1,2,3}$ are given by

$$\sigma_1 = \begin{pmatrix} 0 & 1 \\ 1 & 0 \end{pmatrix}, \quad \sigma_2 = \begin{pmatrix} 0 & -i \\ i & 0 \end{pmatrix}, \quad \sigma_3 = \begin{pmatrix} 1 & 0 \\ 0 & -1 \end{pmatrix}.$$

The explicit expression for the Euclidean gamma matrices is then

$$\gamma_{1,2,3} = \begin{pmatrix} 0 & -i\sigma_{1,2,3} \\ i\sigma_{1,2,3} & 0 \end{pmatrix}, \quad \gamma_4 = \begin{pmatrix} 0 & 1 \\ 1 & 0 \end{pmatrix}, \quad \gamma_5 = \begin{pmatrix} 1 & 0 \\ 0 & -1 \end{pmatrix}.$$

The scalar field φ is a complex doublet which transforms in the fundamental representation with respect to $SU_W(2)$ transformations. It can be easily shown that $\tilde{\varphi}$ also transforms

2.1 The continuum formulation and its symmetries

like a $SU_W(2)$ vector such that the combination $\begin{pmatrix} \bar{t} & \bar{b} \end{pmatrix}_L \cdot \tilde{\varphi}$ is indeed manifestly $SU_W(2)$ invariant.

Besides the global Baryon number conservation and Euclidean invariance, which translates to Poincaré invariance in Minkowski space, the model possess a $SU(2)_W \times U(1)_Y$ symmetry which shall be demonstrated with the help of the Yukawa term involving the scalar and the fermion fields $\begin{pmatrix} \bar{t} & \bar{b} \end{pmatrix}_L \cdot \varphi\, b_R$:

$$\varphi' := V \cdot \varphi \cdot V^{-1} = e^{-i\epsilon^A T^A} \cdot \varphi,$$

$V \in SU(2), T^A : A \in \{1\ldots 3\}$ are generators of $SU(2)$.

$$Q' := V \cdot Q \cdot V^{-1} = e^{-i\epsilon^A T^A} \cdot Q,$$

$$Q := \begin{pmatrix} t_L \\ b_L \end{pmatrix}$$

$$b'_R := b_R$$

$$t'_R := t_R$$

With the above transformation properties it is clear the above Lagrangian is $SU(2)_W$ invariant if the fields in the Lagrangian (2.1) are substituted with the primed fields. It is vital for the $SU_W(2)$ transformation, that it does not affect the right handed fields at all (here: $t'_R = t_r, b'_R = b_r$). The quantum number with respect to $SU_W(2)$ is given by the eigenvalue of the third generator T^3 and is $\frac{1}{2}$ for the upper component and $-\frac{1}{2}$ for the lower component of the $SU_W(2)$ doublet. The transformation properties of $U_Y(1)$ affects both components of the fermion fields and the scalar fields.

The quantum number of the $U_Y(1)$ symmetry has to be chosen such that the electric charges of fermions are reproduced in the broken phase. The electric charges of the quarks are given in fractions such that the observed mesons and baryons e.g. the proton have unit charge. The charge of the top quark is $\frac{2}{3}$ while the charge of the bottom quark is

$-\frac{1}{3}$. The transformation properties of the fields with respect to $U_Y(1)$ are given by

$$\varphi' := U \cdot \varphi \cdot U^{-1} = e^{-i\epsilon Y} \cdot \varphi = e^{-i\frac{\epsilon}{2}} \varphi$$

$$Q'_L := U \cdot Q_L \cdot U^{-1} = e^{-i\epsilon Y} \cdot Q_L = e^{-i\frac{\epsilon}{6}} Q_L$$

$$b'_R := U \cdot b_R \cdot U^{-1} = e^{-i\epsilon Y} \cdot b_R = e^{+i\frac{\epsilon}{3}} b_R$$

$$t'_R := U \cdot t_R \cdot U^{-1} = e^{-i\epsilon Y} \cdot t_R = e^{+i\frac{2}{3}\epsilon} t_R \tag{2.2}$$

With the above rules at hand it is easy to show that the Yukawa terms are also symmetric under $U_Y(1)$

$$\begin{pmatrix} \bar{t}' \\ \bar{b}' \end{pmatrix}_L^T \cdot \varphi' b'_R = \begin{pmatrix} \bar{t} \\ \bar{b} \end{pmatrix}_L^T e^{+i\frac{\epsilon}{6}} \cdot \left(e^{-i\frac{\epsilon}{2}} \varphi\right) \left(e^{+i\frac{\epsilon}{3}} b_R\right)$$

$$= \begin{pmatrix} \bar{t} \\ \bar{b} \end{pmatrix}_L^T \cdot \varphi \, b_R.$$

Within the standard model, there exists a symmetric and a spontaneously broken phase. They are characterized by the value of the scalar vacuum expectation value (*vev*), which vanishes if the ground state respects the symmetry of the Lagrangian and which is non-zero otherwise. The Higgs mechanism exploits the fact that the scalar vacuum expectation value

$$\langle \Omega | \varphi | \Omega \rangle$$

can have a non vanishing value. The above expression for the vacuum expectation value is not an invariant observable under the electroweak symmetry and an infinitesimal symmetry transformation of the scalar field yields

$$\langle \Omega | \varphi' | \Omega \rangle = \langle \Omega | e^{-i\epsilon^A \Gamma^A} \varphi | \Omega \rangle$$
$$= \langle \Omega | \left\{ 1 - i\epsilon^A \Gamma^A \right\} \varphi | \Omega \rangle$$
$$= \langle \Omega | \varphi | \Omega \rangle - i\epsilon \langle \Omega \Gamma^{\dagger A} | \varphi | \Omega \rangle.$$

Here Γ^A stands for an arbitrary generator of some given symmetry. From the above relation one concludes that the scalar vacuum expectation value is zero if

$$\Gamma^A |\Omega\rangle = 0, \quad A \in \{1, \ldots N\}$$

The scalar *vev* is non zero if at least one the generators Γ^A applied to the vacuum state does not vanish, i.e. at least one of the symmetries is broken. The mechanism of broken symmetries allows to define a gauge theory with massless gauge bosons and chirally invariant fermions such that the theory is perturbatively renormalizable. Depending on the vacuum state however, the theory may exhibit a phase where the gauge bosons are massive and the scalar vacuum expectation value generates massive fermions. This phase corresponds to the phenomenologically observed massive $W^\pm$ and Z bosons and massive fermions. The Goldstone theorem states that for each broken generator of a continuous global symmetry, the spectrum of the theory contains a massless scalar particle, the Goldstone boson. In the case considered here, the electroweak symmetry algebra contains four generators, the identity and the three Pauli matrices. As will be shown below, the scalar vev is invariant under a $U_{em}(1)$ symmetry and thus three Goldstone bosons are then expected in the spectrum of the theory. As the scalar doublet consists of four real valued scalar fields, there is one massive scalar particle left, which will be denoted as the Higgs boson. The hypercharge is chosen such that the Higgs boson is neutral, i.e. the quantum number of the Higgs field with respect to $U_{em}(1)$ transformations is zero.

In order to keep the arguments as transparent as possible, it will be assumed that the vev has the following form

$$\varphi_0 := \langle \Omega | \varphi | \Omega \rangle = \begin{pmatrix} 0 \\ \frac{v}{\sqrt{2}} \end{pmatrix}, \quad v = const.$$

Clearly if $|v| > 0$, it is not invariant under general $SU_W(2) \times U_Y(1)$ transformations but it does not break all symmetries. Choosing a subgroup of $SU_W(2) \times U_Y(1)$ with $\epsilon^1 = \epsilon^2 = 0$ and $\epsilon^3 = \epsilon$ yields

$$\varphi_0' = e^{-iT^A \epsilon^A} e^{-i\epsilon Y} \varphi_0,$$
$$\Rightarrow \varphi_0' = e^{-i(T^3 + Y)\epsilon} \varphi_0, \quad T^3 \varphi_0 = -\frac{1}{2}\varphi_0, \quad Y\varphi_0 = \frac{1}{2}\varphi_0$$
$$= \varphi_0.$$

Hence, the combination of the generators

$$Q = T^3 + Y$$

gives an unbroken $U(1)$ symmetry and its eigenvalues will be identified with the electric charge. The hyper-charges in equation (2.2) are chosen such that eigenvalues of Q yield the correct electric charges for the neutral Higgs boson and the charged quarks.

The path integral is the basic quantity in order to define the Euclidean quantum field theory

$$\mathcal{F}[\mathcal{O}] := \mathcal{N} \int \underbrace{\prod_{x \in \mathbb{R}^4} \mathrm{d}\varphi(x)}_{D\varphi} \underbrace{\prod_{x \in \mathbb{R}^4} \mathrm{d}\varphi^\dagger(x)}_{D\varphi^\dagger} \underbrace{\prod_{x \in \mathbb{R}^4} \mathrm{d}\psi(x) \prod_{x \in \mathbb{R}^4} \mathrm{d}\overline{\psi}(x)}_{D\psi D\overline{\psi}}$$
$$\times \mathcal{O}\left(\varphi, \varphi^\dagger, \psi, \overline{\psi}\right) e^{-S_E\left(\varphi, \varphi^\dagger, \psi, \overline{\psi}\right)}. \quad (2.3)$$

The fermionic variables ψ and $\overline{\psi}$ collect the top and bottom quark fields

$$\psi := \begin{pmatrix} t \\ b \end{pmatrix}, \quad \overline{\psi} := \begin{pmatrix} \overline{t} \\ \overline{b} \end{pmatrix}.$$

The Normalization $\mathcal{N}$ is chosen such that $\mathcal{F}[\mathcal{O}]$ evaluates to unity for $\mathcal{O} \equiv 1$

$$\Rightarrow \mathcal{N} = \int D\varphi D\varphi^\dagger D\psi\, D\overline{\psi}\, e^{-S_E\left(\varphi, \varphi^\dagger, \psi, \overline{\psi}\right)}.$$

The expectation value of observables are then defined by the path integral

$$\langle \mathcal{O}\left(\varphi, \varphi^\dagger, \psi, \overline{\psi}\right) \rangle = \mathcal{F}[\mathcal{O}].$$

There are various methods to evaluate the path integral, which can roughly be classified by those which rely on perturbative expansions that have to be truncated in order to be calculable and those which are based on a discretized and finite space time lattices. The latter allows for both, perturbative and non-perturbative evaluations of the path integral. The non-perturbative calculations rely heavily on numerical simulations of the model. Though the validity of perturbation theory depends on the values of the (renormalized) couplings of the model, perturbation theory turned out to be extremely useful in its predictive power. Within the non-perturbative framework of lattice field theory the model can be evaluated at any point of the bare parameter space and is capable to reveal eventual non perturbative effects based on first principles.

2.2 The model on a discretized space-time lattice

This work relies on a discretized Euclidean space time lattice in order to evaluate the path integral. While the discretization of the scalar field can easily be performed by

2.2 The model on a discretized space-time lattice

substituting the derivatives with the lattice nearest neighbor coupling, the fermion sector is known to suffer from conceptual difficulties.

The scalar lattice action and some notations will be defined below. A short overview on lattice chiral symmetry is discussed afterwards. It is common in lattice field theory to rewrite the scalar sector by rescaling the fields with a factor $\sqrt{2\kappa}$ and furthermore, the scalar doublet can be expressed as a quarternion

$$\phi := \begin{pmatrix} \tilde{\varphi}_1 & \varphi_1 \\ \tilde{\varphi}_2 & \varphi_2 \end{pmatrix} =: \phi^0 \mathbb{I} - i\sigma^j \phi^j, \quad \phi_\mu \in \mathbb{R}.$$

It will turn out to be useful to introduce the following notations

$$\theta_\mu := (\mathbb{I}, -i\vec{\tau}), \quad \bar{\theta}_\mu := (\mathbb{I}, +i\vec{\tau})$$

$$\Rightarrow \phi = \phi^\mu \theta_\mu, \qquad \phi^\dagger = \phi^\mu \bar{\theta}_\mu.$$

The rescaled fields are:

$$\Phi^\mu := \frac{1}{\sqrt{2\kappa}} \phi^\mu.$$

Finally the scalar fields in the usual notation (2.1) can be recovered by identifying

$$\begin{pmatrix} \varphi_1 \\ \varphi_2 \end{pmatrix} = \sqrt{2\kappa} \begin{pmatrix} \Phi_x^2 + i\Phi_x^1 \\ \Phi_x^0 - i\Phi_x^3 \end{pmatrix}.$$

Based on the scalar part in the Lagrangian (2.1), the scalar lattice action is derived by replacing the integral with a finite sum and the derivatives with finite differences. The detailed derivation is shifted to the appendix A. The scalar lattice action is given by

$$S_\Phi = -\kappa \sum_{x,\mu} \Phi_x^\dagger (\Phi_{x+\mu} + \Phi_{x-\mu}) + \sum_x \Phi_x^\dagger \Phi_x + \hat{\lambda} \sum_x \left(\Phi_x^\dagger \Phi_x - N_f \right)^2.$$

The lattice spacing a is set to unity. As mentioned before, the couplings are scaled by the κ parameter. The parameterization of the Lagrangian in (2.1) can by recovered with the identities:

$$\lambda = \frac{\hat{\lambda}}{4\kappa^2}, \quad m_0^2 = \frac{1 - 2N_f \hat{\lambda} - 8\kappa}{\kappa}, \quad y_{t,b} = \frac{\hat{y}_{t,b}}{\sqrt{2\kappa}}.$$

Lattice Higgs-Yukawa models aim to investigate the spontaneous breakdown of the $SU(2)_W \times U_Y(1)$ symmetry into a remaining $U(1)_{em}$ symmetry. While in the symmetric phase all considered fermions are exactly massless and the particle spectrum reflects the underlying symmetry of the Lagrangian, the broken phase (also known as the Higgs phase) contains massive fermions and reveals a non-degenerate spectrum of scalar particles. The phase structure of the considered Higgs-Yukawa model has been explored in a series of papers [27, 28, 29, 26].

As was claimed at the beginning of this chapter, chiral symmetry is of primal conceptual importance for the model. Chiral symmetry can be established with the relation

$$\gamma_5 D + D\gamma_5 = 0.$$

Henceforth, the above relation will be denoted as the continuum chiral symmetry. The continuum chiral symmetry forbids a fermion mass term in the Dirac operator. An explicit mass term, which mixes right and left handed components of the fermion spinor, is also not compatible with the electroweak $SU_W(2)$ symmetry because it treats left handed components different than right handed components. In the continuum theory the projectors $P_\pm$

$$P_\pm := \frac{1}{2}(1 \pm \gamma_5), \qquad P_\pm^2 = P_\pm,$$
$$\mathbb{I} = P_+ + P_-, \qquad P_\pm P_\mp = 0$$

are used to define the left and right handed components

$$\psi_L := P_-\psi, \quad \psi_R := P_+\psi.$$

Given the exact continuum chiral symmetry and the above projectors, the free part of the Lagrangian (2.1) can be written as a sum of right and left handed spinor fields

$$\overline{\psi}\slashed{D}\psi = \overline{\psi}_R\slashed{D}\psi_R + \overline{\psi}_L\slashed{D}\psi_L.$$

Here the relation $P_\pm D = DP_\mp$ was used, which makes explicit use of chiral symmetry. An explicit mass term mixes both chiral components

$$\overline{\psi}\psi = \overline{\psi}_L\psi_R + \overline{\psi}_R\psi_L$$

and hence it is not invariant under $SU_W(2)$ transformations as it violates chiral symmetry.

Establishing chiral symmetry on the lattice is more complicated and was a long lasting challenge. The main conflict in the lattice formulation of massless fermions is phrased in the Nielson-Ninomiya theorem. The theorem states that there is no lattice Dirac operator such that the fermion action simultaneously fulfils the following conditions [50]:

2.2 The model on a discretized space-time lattice

1. chiral symmetry

2. describes a single physical fermion

3. locality

4. invariance under translations

In 1982 Ginsparg and Wilson [34] proposed a relation which defines a class of lattice Dirac operators and which is since then known as the Ginsparg-Wilson relation

$$\gamma_5 D + D\gamma_5 = aD\gamma_5 RD. \tag{2.4}$$

a is the lattice spacing and R is a positive constant in momentum space. The Ginsparg-Wilson relation can be utilized to construct a lattice modified chiral symmetry which recovers the desired continuum symmetry in the limit $a \to 0$. In order to define chiral symmetry on the lattice, it is therefore necessary to modify the projectors $P_\pm$ such that they take the Ginsparg-Wilson type chiral symmetry into account. Starting from the Ginsparg-Wilson relation, this can be easily seen by

$$\gamma_5 D + D\gamma_5 = aDR\gamma_5 D$$
$$\Rightarrow \gamma_5 D + D\gamma_5 (1 - aRD) = 0.$$

The lattice modified lattice projectors are then given by

$$\hat{P}_\pm := \frac{1}{2}(1 \pm \hat{\gamma}_5), \quad \hat{\gamma}_5 := \gamma_5(1 - aRD).$$

$\hat{P}_\pm$ are indeed projectors; some relations involving the lattice projectors are summarized below

$$\hat{P}_\pm + \hat{P}_\mp = \mathbb{I}, \quad \hat{P}_\pm^2 = \hat{P}_\pm, \quad \hat{P}_\pm \hat{P}_\mp = 0, \quad \hat{\gamma}_5^2 = \mathbb{I}, \quad \hat{\gamma}_5 \hat{P}_\pm = \pm \hat{P}_\pm.$$

The modified projectors are used to define the chiral components of the spinor fields on the lattice

$$P_\pm D = \frac{1}{2}(1 \pm \gamma_5) D \qquad \text{using eq. (2.4)}$$
$$= \frac{1}{2} D (1 \mp \gamma_5 (1 - aRD)) = D\hat{P}_\mp.$$

Finally the left and right handed spinor fields on the lattice are defined by

$$\psi_R = \hat{P}_+ \psi, \qquad \psi_L = \hat{P}_- \psi,$$
$$\overline{\psi}_R = \overline{\psi} P_-, \qquad \overline{\psi}_L = \overline{\psi} P_+.$$

As in the continuum theory the free part of the fermion Lagrangian can be written as sum of the left and right handed lattice spinor fields

$$\overline{\psi}D\psi = \overline{\psi}_L D\psi_L + \overline{\psi}_R D\psi_R + \underbrace{\overline{\psi}_L D\psi_R + \overline{\psi}_R D\psi_L}_{=0}.$$

It is easily shown that the last term vanishes

$$\overline{\psi}_L D\psi_R = \overline{\psi} \hat{P}_R D \hat{P}_R \psi = \overline{\psi} D \hat{P}_L \hat{P}_R \psi = 0.$$

The Ginsparg-Wilson relation allows to *define* the left and right handed components of the lattice spinor fields and ensures that there is no mixture between these components even for finite lattice spacing a. The lattice modified projection operators are vital to the $SU_W(2)$ transformations as they allow to perform transformations which only affect the left handed component of the spinor fields.

In order to define the lattice action, a Ginsparg-Wilson type Dirac operator has to be introduced. The here presented results are based on the Neuberger overlap operator [51], which satisfies the Ginsparg-Wilson relation and is given by

$$\mathcal{D}^{(ov)} = \frac{\rho}{a}\left\{1 + \frac{A}{\sqrt{A^\dagger A}}\right\}, \quad A = D^W - \frac{\rho}{a}, \quad 0 < \rho < 2r \qquad (2.5)$$

ρ is chosen to be $\frac{1}{R}$. D^W is the Wilson Dirac operator which lifts the unwanted doublers but it does not fulfil the Ginsparg-Wilson relation (2.4). The operator will be constructed from its eigenvalues in momentum space. The eigenvectors and eigenvalues of the doublet operator $\mathcal{D}^{(ov)}\mathbb{I}$ is summarized:

$$\Psi_x^{p,\zeta\epsilon k} = e^{ip\cdot x} \cdot u^{\zeta\epsilon k}(p), \quad u^{\zeta\epsilon k}(p) = \sqrt{\frac{1}{2}}\begin{pmatrix} u^{\epsilon k}(p) \\ \zeta u^{\epsilon k}(p) \end{pmatrix}, \quad \zeta = \pm 1, \epsilon = \pm 1, k \in \{1,2\}$$

$$u^{\epsilon k}(p) = \sqrt{\frac{1}{2}}\begin{pmatrix} \xi_k \\ \epsilon \frac{\tilde{p}\theta}{\sqrt{\tilde{p}^2}}\xi_k \end{pmatrix} \text{ for } \tilde{p} \neq 0 \quad \text{and} \quad u^{\epsilon k}(p) = \sqrt{\frac{1}{2}}\begin{pmatrix} \xi_k \\ \epsilon \xi_k \end{pmatrix} \text{ for } \tilde{p} = 0,$$

$$\theta = (\mathbb{I}, -i\vec{\tau}),$$
$$\overline{\theta} = (\mathbb{I}, +i\vec{\tau}).$$

$u^{\epsilon k}$ is the usual four component spinor and Ψ are the eigenvectors. Finally the corresponding eigenvalues are

$$\nu^\epsilon(p) = \frac{\rho}{a} + \frac{\rho}{a} \cdot \frac{\epsilon i \sqrt{\tilde{p}^2} + a\frac{r}{2}\hat{p}^2 - \frac{\rho}{a}}{\sqrt{\tilde{p}^2 + (a\frac{r}{2}\hat{p}^2 - \frac{\rho}{a})^2}}.$$

2.2 The model on a discretized space-time lattice

The momenta $\hat{p}$ and $\tilde{p}$ are defined by the discretized lattice momenta

$$\hat{p}_\mu = 2a \sin\left(\frac{\pi}{L_\mu} n_\mu\right), \quad n_\mu \in \{0, \ldots, L_\mu - 1\}$$

$$\tilde{p}_\mu = a \sin\left(\frac{2\pi}{L_\mu} n_\mu\right), \quad n_\mu \in \{0, \ldots, L_\mu - 1\}.$$

In the above relation there shall be no sum over repeated indices.

The Neuberger overlap operator respects all conditions listed as prerequisites of the Nielson-Ninomiya theorem except the continuum chiral symmetry and thus it is not a contradiction to the above no-go theorem. The locality of the Neuberger operator is not obvious but it has been shown in [39] that it is indeed local in the sense that $\mathcal{D}^{(ov)}_{xy}$ exhibits an exponential decay with respect to the spatial distance $|x - y|$.

Given a Dirac operator satisfying the Ginsparg-Wilson relation, it is possible to translate the chirally invariant Higgs-Yukawa model on a finite discretized space time lattice. An action involving the lattice spinor fields was proposed in [47] where auxiliary fields were introduced in order to keep the transformations of the spinors as in the continuum formulation. These auxiliary fields though, do not propagate and can be eliminated. The transformation properties then involve the modified projection operators and will be given below. The lattice fermion action is given by

$$S_F = \sum_{x,y,\alpha,\beta} \begin{pmatrix} \bar{t}^\alpha_x \\ \bar{b}^\alpha_x \end{pmatrix} \mathbb{I}_2 \left\{ D^{\alpha\beta}_{x,y} + \hat{y} \left(P_- \phi \hat{P}_- + P_+ \phi^\dagger \hat{P}_+ \right)^{\alpha\beta}_{x,y} \right\} \begin{pmatrix} t^\beta_y \\ b^\beta_y \end{pmatrix}.$$

The fermion matrix which will be used more often throughout this work is defined by

$$\mathcal{M}^{\alpha\beta}_{xy} = \left(\mathcal{D}^{(ov)}\right)^{\alpha\beta}_{xy} + \hat{y}\,\Phi^\mu_x \left(P^{\alpha\beta}_+ \theta^\dagger_\mu + P^{\alpha\beta}_- \theta_\mu \right) \underbrace{\left(1 - \frac{1}{2}aR\left(\mathcal{D}^{(ov)}\right)^{\alpha\beta}_{xy}\right)}_{P_\pm \hat{P}_\pm}, \tag{2.6}$$

$$\theta = (\mathbb{I}, -i\vec{\tau}),$$

$$\theta^\dagger = (\mathbb{I}, +i\vec{\tau}).$$

While the free-fermion Neuberger Dirac operator can be analytically constructed from its eigenvectors and eigenvalues in momentum space, the coupling to the scalar field in position space prohibits such an approach. In the simulations it becomes therefore necessary to perform fast Fourier transformations of the scalar fields. The fast Fourier transformation is known to be of order $N \log(N)$ where N is the length of the vector to be transformed. The complex scalar field has in general four degrees of freedom and thus N is identical to $4TL^3$. Nevertheless, it is the fast discrete Fourier transformation, which allows to evaluate the fermionic action efficiently. The largest lattice volumes, which could be simulated with the above model, were 40^4. On a modern computing centre such as the "Norddeutscher Verbund für Hoch- und Höchstleistungsrechnen" (HLRN), about 40 configurations could be produced per day. The analysis of scattering phases which will be discussed in Chapter 4 needs large lattices in order to probe the system below the inelastic threshold of the two Goldstone system. Sophisticated preconditioning of the fermion matrix as well as adequate computing resources are inevitable in order to extract physical quantities.

Finally, after the conceptual properties of the electroweak symmetry on the lattice has been elaborated, the symmetries of the lattice action can be summarized. The Euclidean quantum field theory is manifestly invariant under global $O(4)$ symmetry and translations. Both reflect the Poincaré symmetry in Minkowski space. Furthermore, finite volume and the boundary conditions reduce the continuous $O(4)$ symmetry to the cubic symmetry. All physical quantities in this work are obtained after an extrapolation to infinite volume. Additionally, the action is invariant under global $SU_W(2) \times U_Y(1)$ transformations. Contrary to the continuum transformation rules, special care is needed where the lattice chiral projectors are used. The chiral components of the adjoint fields $\overline{\psi}$ are projected with the usual projector $P_\pm$ while the fields ψ are transformed with the lattice modified

2.2 The model on a discretized space-time lattice

operator $\hat{P}_\pm$. The transformation properties are:

SU$_W$(2) transformations:

$$\phi' := e^{-i\epsilon^A T^A}\phi$$

$$\phi'^\dagger := \phi^\dagger e^{+i\epsilon^A T^{A\dagger}}$$

$$Q'_L = \hat{P}_- \begin{pmatrix} t \\ b \end{pmatrix}' := e^{-i\epsilon^A T^A}\hat{P}_- \begin{pmatrix} t \\ b \end{pmatrix}$$

$$\overline{Q}'_L = \begin{pmatrix} \bar{t} \\ \bar{b} \end{pmatrix}' P_+ := \begin{pmatrix} \bar{t} \\ \bar{b} \end{pmatrix} P_+ e^{+i\epsilon^A T^{A\dagger}}$$

$$t'_R = \hat{P}_+ t' := \hat{P}_+ t = t_R$$

$$\bar{t}'_R = \bar{t}' P_- := \bar{t} P_- = \bar{t}_R$$

$$b'_R = \hat{P}_+ b' := \hat{P}_+ b = b_R$$

$$\bar{b}'_R = \bar{b}' P_- := \bar{b} P_- = \bar{b}_R$$

U$_Y$(1) transformations:

$$\phi' := e^{-i\frac{\epsilon}{2}}\phi$$

$$\phi'^\dagger := \phi^\dagger e^{+i\frac{\epsilon}{2}}$$

$$Q'_L := e^{-i\frac{\epsilon}{6}}\hat{P}_- \begin{pmatrix} t \\ b \end{pmatrix}$$

$$\overline{Q}'_L := \begin{pmatrix} \bar{t} \\ \bar{b} \end{pmatrix} P_+ e^{+i\frac{\epsilon}{6}}$$

$$t'_R := e^{-i\frac{2}{3}\epsilon}\hat{P}_+ t$$

$$\bar{t}'_R := \bar{t} P_- e^{+i\frac{2}{3}\epsilon}$$

$$b'_R := e^{+i\frac{1}{3}\epsilon}\hat{P}_+ b$$

$$\bar{b}'_R := \bar{b} P_- e^{-i\frac{1}{3}\epsilon}$$

T^A, $A \in \{1,\ldots,3\}$ are the generators of $SU_W(2)$ and fulfil the $SU(2)$ algebra. The hyper charges Y are the same as in the continuum theory and are taken from (2.2).

The full Euclidean discretized action defined in given by

$$S = -\kappa \sum_{x,\mu} \Phi^\dagger_x (\Phi_{x+\mu} + \Phi_{x-\mu}) + \sum_x \Phi^\dagger_x \Phi_x + \hat{\lambda} \sum_x \left(\Phi^\dagger_x \Phi_x - N_f\right)^2$$

$$+ \sum_{x,y} \begin{pmatrix} \bar{t}^\alpha_x \\ \bar{b}^\alpha_x \end{pmatrix} \left\{ \mathbb{I}_2 D^{\alpha\beta}_{x,y} + \hat{y}\left(P_-\phi\hat{P}_- + P_+\phi^\dagger\hat{P}_+\right)^{\alpha\beta}_{x,y} \right\} \begin{pmatrix} t^\beta_y \\ b^\beta_y \end{pmatrix}. \quad (2.7)$$

The above defined lattice Higgs-Yukawa model has recently been studied in various context relevant for phenomenology. The phase structure of the model has been analyzed with the help of a large N expansion and was confronted with numerical data [27, 28, 29, 26]. Due to the triviality of the theory, the maximal and minimal Higgs boson mass can be determined in dependence of the cut off of the theory. The procedure is explained in the next section. The final results on the mass bounds of the Higgs boson are published in [31, 30]. Recently, the interest in the existence of a heavy fourth generation of fermions has been renewed [41, 40, 21, 14]. Heavy fermions can alter the afore mentioned mass bounds and may reveal non perturbative effects. The upper bound for the fermion mass, which can be generated via Yukawa couplings, is constrained to be around 550 GeV. This constraint was computed in perturbation theory and relies on partial wave unitarity. The Higgs boson mass bounds were reconsidered in a scenario, where the heavy fourth generation top quark has a mass of around 700 GeV. These results were published in [32, 33].

2.3 Simulation strategy

The simulation algorithm is a polynomial hybrid Monte Carlo (pHMC) algorithm which incorporates dynamical overlap fermions. The polynomial algorithm and its benefits are described in [22, 23, 24] and an implementation of the algorithm and improvements are given in [25]. Some aspects of the algorithm are sketched here in order to explain the chosen parameters of the simulation algorithm.

The path integral was given in (2.3) and is repeated here

$$\mathcal{F}[\mathcal{O}] := \mathcal{N} \int D\Phi D\psi D\overline{\psi} \, \mathcal{O}\left(\Phi, \psi, \overline{\psi}\right) e^{-S\left(\Phi, \psi, \overline{\psi}\right)}.$$

The integral over the Grassmann fields $t, \overline{t}, b, \overline{b}$ can be expressed in a determinant

$$\mathcal{F}[\mathcal{O}] := \mathcal{N} \int D\Phi \, \det(\mathcal{M}) \, \mathcal{O}\left(\Phi, \psi, \overline{\psi}\right) e^{-S_\Phi(\Phi)}.$$

where $\mathcal{M}$ is given in (2.6). Any hermitian, positive definite matrix A can be rewritten with the help of a Gaussian integral

$$\det A = \int d\omega d\omega^\dagger \, e^{-\frac{1}{2}\omega^\dagger A^{-1}\omega}.$$

Computing the determinant of a large matrix like $\mathcal{M}$ is a numerically demanding task and hard to manage for large lattice volumes. It is therefore a convenient alternative to compute the Gaussian integral up a desired precision. The fermion matrix $\mathcal{M}$ is neither positive nor is it hermitian. To circumvent this issue one considers the matrix $\mathcal{M}\mathcal{M}^\dagger$, which per construction fulfils the necessary properties. The path integral can now be written as

$$\mathcal{F}[\mathcal{O}] := \mathcal{N} \int D\Phi D\omega D\omega^\dagger \, \text{sgn}(\mathcal{M}) \, \mathcal{O}\left(\Phi, \psi, \overline{\psi}\right) e^{-S_\Phi(\Phi) - S_F(\omega, \omega^\dagger)}$$
$$S_F\left(\omega, \omega^\dagger\right) := \frac{1}{2}\omega^\dagger \left(\mathcal{M}\mathcal{M}^\dagger\right)^{-\frac{1}{2}} \omega$$

2.3 Simulation strategy

It can be shown that the fermion determinant is real and thus the sign of the determinant has to be determined when it is computed with the help of a Gaussian integral. $\text{sgn}(\mathcal{M})$ in the above equation stands for the sign of the determinant of the fermion matrix.

In order to show this, it is sufficient to construct an operator V such that

$$V \mathcal{M} V^\dagger = \mathcal{M}^*.$$

Given such an operator, one can show that if λ is an eigenvalue with eigenvector η of the fermion matrix $\mathcal{M}$, then λ^* is an eigenvalue with eigenvector $\xi^* = V^T \eta^*$. The above statement is easy to verify

$$\mathcal{M} \cdot \eta = \lambda \eta$$
$$\Rightarrow \eta^\dagger \cdot \underbrace{V V^{-1}}_{=\mathbb{I}} \mathcal{M}^\dagger V = \lambda^* \eta^\dagger \cdot V$$
$$\left(V^\dagger \cdot \eta\right)^\dagger V^{-1} \mathcal{M}^\dagger V = \lambda^* \left(V^\dagger \cdot \eta\right)^\dagger$$
$$\xi := V^\dagger \eta$$
$$\Rightarrow \xi^\dagger \mathcal{M}^T = \lambda^* \xi^\dagger$$
$$\Rightarrow \mathcal{M} \cdot \xi^* = \lambda^* \xi^*.$$

The operator V is given by

$$V := \sigma_2 C \gamma_5, \qquad C := \gamma_4 \gamma_2.$$

The part of the fermion matrix can be collected within a matrix B

$$B^{\alpha\beta}_{xy} := \Phi^\mu_x \left(P^{\alpha\beta}_+ \theta^\dagger_\mu + P^{\alpha\beta}_- \theta_\mu \right)$$
$$\Rightarrow \mathcal{M}^{\alpha\beta}_{xy} = \left(\mathcal{D}^{(ov)}\right)^{\alpha\beta}_{xy} + \hat{y} B^{\alpha\beta}_{xy} \left(1 - \frac{1}{2} a R \left(\mathcal{D}^{(ov)}\right)^{\alpha\beta}_{xy}\right).$$

The Pauli matrix σ_2 does not affect the Wilson operator and transforms only the operator B. At the same time the charge conjugation C and the operator γ_5 does not alter B but the Wilson operator

$$\sigma_2 B \sigma_2 = B^*$$
$$C \gamma_5 \, \mathcal{D}^W \gamma_5 C^\dagger = \left(\mathcal{D}^W\right)^*$$

The overlap operator is a power series in $\mathcal{D}^W$ and the above properties directly apply to the overlap operator. Hence, the operator V transforms the fermion matrix to its complex conjugate and ensures that the determinant of the fermion matrix is strictly real valued. The latter follows from

$$0 = \det\left(\mathcal{M} - \lambda \mathbb{I}\right) = \det\left\{V\left(\mathcal{M} - \lambda \mathbb{I}\right)V^\dagger\right\} = \det\left(\mathcal{M}^* - \lambda \mathbb{I}\right)$$
$$\Rightarrow \det\left(\mathcal{M} - \lambda^* \mathbb{I}\right) = 0.$$

The inverse square root of the fermion matrix is computed via a polynom of finite degree

$$P\left(\mathcal{MM}^\dagger\right) = \left(\mathcal{MM}^\dagger\right)^{-\frac{1}{2}}.$$

As the polynom is truncated at a finite degree, a systematic error is introduced, which however, can be avoided by introducing a weight factor which corrects for potential lack of precision in the polynomial approximation. The weight factor in the algorithm used here is defined by

$$S_F\left(\omega, \omega^\dagger\right) := \frac{1}{2}\omega^\dagger \mathcal{MM}^{\dagger -\frac{1}{2}}\omega$$
$$= \frac{1}{2}\omega^\dagger\left\{\left(\mathcal{MM}^\dagger\right)^{-\frac{1}{2}} + P\left(\mathcal{MM}^\dagger\right) - P\left(\mathcal{MM}^\dagger\right)\right\}\omega$$
$$\Rightarrow W\left(\Phi, \omega^\dagger, \omega\right) := e^{-\frac{1}{2}\omega^\dagger\left(\left(\mathcal{MM}^\dagger\right)^{-\frac{1}{2}} - P(\mathcal{MM}^\dagger)\right)\omega}. \tag{2.8}$$

After separating the weight factor from the fermionic action, the expectation value of observables is given by

$$\mathcal{F}[\mathcal{O}] = \mathcal{N}\int D\Phi D\omega D\omega^\dagger \operatorname{sgn}(\mathcal{M})$$
$$\mathcal{O}\left(\Phi, \psi, \overline{\psi}\right) W\left(\Phi, \omega^\dagger, \omega\right) e^{-S_\Phi(\Phi) - \frac{1}{2}\omega^\dagger P(\mathcal{MM}^\dagger)\omega}.$$

There are different ways to treat the weight factor. Here, the weight factor is computed during the simulation for each successive configuration Φ. Computing the weight factor during the simulation has the advantage that the quantity $\omega^\dagger P\left(\mathcal{MM}^\dagger\right)\omega$ has to be evaluated anyway as it appears in the fermionic action. The only additional cost is therefore to compute the inverse square root of the matrix $\mathcal{MM}^\dagger$, which is done up to machine precision with a Lanczos based method. A detailed overview of the algorithm and its extension to treat $\left(\mathcal{MM}^\dagger\right)^{-\alpha}$ for any real number α is given in [25].

The basic principle of the Monte Carlo integration method is to substitute the high dimensional integral over a domain D by a so-called importance sampling. In this method,

2.3 Simulation strategy

the measure is changed according to the probability distribution of the integrand. If for instance only the the scalar field is considered, this amounts to sample field configurations with the weight $\mathcal{N} \, e^{-S_\Phi}$. The simulation algorithm provides a set of successive field configurations and the expectation value of observables are given by averaging over the field configurations

$$\langle \mathcal{O} \rangle = \lim_{N \to \infty} \frac{1}{N} \sum_{i=1}^{N} \mathcal{O}(\Phi_i) \, \text{sgn}(\mathcal{M}).$$

Details on the algorithm and improvements of the algorithm including the Fourier acceleration to reduce the auto correlation times and the preconditioning techniques to reduce the condition number of the fermion matrix $\mathcal{M}\mathcal{M}^\dagger$ are given in the references at the beginning of this section.

The polynomial approximation used to compute the inverse square root of the fermion matrix induces mainly two technical parameters which have to be tuned depending on the chosen bare parameters of the theory. The polynom is truncated at a finite degree and it is defined on a finite interval $[\epsilon, \lambda]$. The interval is always scaled with λ such that the upper bound of the interval is identical to one ($\lambda = 1$). The lower bound ϵ depends strongly on the parameter space. It has to be tuned such that only a minor fraction of the eigenvalues of the fermion matrix lay below this bound. The eigenvalues below ϵ are then taken into account exactly by the weight factor. The degree of the polynom determines the cost of the simulation, i.e. the lower the degree the lower the cost. However, the degree determines also the magnitude of the fluctuation of the weight factor, i.e. the higher the degree the lower the fluctuations. A highly fluctuating weight factor may cause large statistical errors in the considered observable and thus, one has to find a balance between the fluctuation of the weight factor and the polynom degree for the special case of interest.

The rest of this section deals with the definition of observables which determine the phase structure of the model and the strategy to extract the masses of the particles in the theory.

The bare parameters of the theory are λ_0, m_0 and y_0 (or, equivalently $\kappa, \hat{\lambda}, \hat{y}$). The subscript zero denotes that all considered parameters are not renormalized. The observables which are evaluated within this work and the addressed questions focuses on the broken phase of the model. The parameter κ respectively m_0 has thus to be chosen such that the simulation point is above the phase transition line. Furthermore the obtained non zero magnetization, which indicates the broken phase, is scaled such that the scalar *vev* meets the phenomenologically known value of 246 GeV. The latter scale is also used to determine the cut off (Λ) of the theory. Within the broken phase, the parameter κ or m_0 is tuned to achieve the desired value of the cut off. Calculations performed within the electroweak standard model imply top quark masses of around 175 GeV. The bare Yukawa coupling y_0 is tuned such that the physical value of the simulated top quark mass is reached. In Chapter 5 effects of higher top quark masses are investigated. The bare Yukawa coupling is then free to take any value, but the specific phenomenological questions addressed there restraints the range of desired top quark masses and accordingly determines the bare Yukawa coupling. Finally, there is one bare parameter λ_0 left unspecified in the model. Due to the fact that the Higgs boson has never been observed in an experiment, its mass and its scattering properties are unknown. Hence, there is no constraint on the bare quartic coupling. It is however known that the renormalized quartic coupling is bounded [48]. The renormalized quartic coupling can never be smaller than zero because in that case, the model does not possess a minimum. The triviality of the model implies that at arbitrary large values of the cut off, the renormalized quartic coupling necessarily needs to vanish. Hence, at any given finite value of the cut off, there is a maximal value of the renormalized quartic coupling. The highest possible Higgs boson mass within this model is thus reached at the above mentioned maximal value of the renormalized quartic coupling. It has been shown in [25] that the lowest Higgs boson mass is reached at vanishing bare quartic coupling λ_0 and correspondingly the largest renormalized quartic coupling is reached at infinite bare quartic coupling.

The symmetric and the broken phase of the model are distinguished by the magnetization which is zero in the symmetric phase and greater than zero in the broken phase. The magnetization is given by

$$\text{mag} := |\overline{\Phi}| = \left(\sum_\alpha \overline{\Phi}_\alpha^2\right)^{\frac{1}{2}}, \qquad \overline{\Phi}_\alpha = \frac{1}{V}\sum_{x \in \mathbb{Z}_L^4} \Phi_\alpha(x). \qquad (2.9)$$

V is the number of discrete space time points $V = L^3 T$. The bare scalar *vev* is obtained after the magnetization is rescaled with the factor $\sqrt{2\kappa}$

$$v := \sqrt{2\kappa}\,\text{mag}\,. \qquad (2.10)$$

2.3 Simulation strategy

The renormalized value of the scalar vev is then

$$v_R := \sqrt{2\kappa}\frac{\overline{\text{mag}}}{\sqrt{Z_G}} \qquad (2.11)$$

where Z_G is the field renormalization factor of the Goldstone fields. The straight forward definition of the scalar vacuum expectation value $\langle\Phi\rangle$ is not invariant under the symmetries of the model and thus the ensemble average necessarily vanishes. It was argued before that the special form of the vacuum expectation value φ_0 respects the $U_{em}(1)$ symmetry. There it was assumed that one of the degenerate vacua is chosen. The Monte Carlo algorithm however, does not a pick a certain ground state, instead it averages over all degenerate vacua. The usual procedure in context of the path integral formulation is to add an external current which couples to one of the scalar field components

$$S_\Phi[J] := S_\Phi + J \sum_{x \in \mathbb{Z}_L^4} \Phi_x^0.$$

This current breaks the symmetry explicitly and expectation values of observables are a functional of the external current J. The physical result is then obtained after to limiting procedures. First one has to take the infinite volume limit at fixed external current J and subsequently the limit for vanishing external current has to be taken

$$\langle \mathcal{O} \rangle = \lim_{J \to 0} \left\{ \lim_{V \to \infty} \langle \mathcal{O}[J] \rangle \right\}.$$

The above limiting procedure implies an enormous numerical task. It was shown in [37, 36] that the definition (5.5) of the scalar vev converges to the vev obtained after taking the twofold limit of $V \to \infty$ and $J \to 0$. The vev is taken according to the definition (5.5) throughout this work unless otherwise stated.

In order to compute the renormalized vev_R one needs the renormalization factor Z of the scalar field. It is known that the Higgs boson field renormalization factor and the Goldstone field renormalization factor are very close to each other. Here the latter will be used in order to determine the renormalized vev_R. In the action defined in (2.7) the value of the lattice spacing has been set to unity. Hence, all obtained lattice results can only be given in units of the lattice spacing and consequently a conversion of the lattice spacing into physical units is needed. The scalar vev is a dimensionful quantity and therefore the renormalized vev_R expressed in lattice units is given by

$$\frac{vev_R}{a} := \frac{1}{\sqrt{Z_G}}\frac{vev}{a}.$$

The physical value of the renormalized *vev* is known to be around 246 GeV and thus the above equation can be used to set the lattice spacing a

$$\Lambda := \frac{1}{a} = \frac{\sqrt{Z_G}}{vev} 246 \text{ GeV}.$$

Λ is the cut off the theory which is here defined to be the inverse lattice spacing. There is no unique definition of the cut off and it rather represents a scale near which the results will strongly depend on the lattice regulator.

The work presented here follows roughly two strategies. In the first scenario one aims to explore the maximum and the minimum attainable Higgs boson masses at varying cut off values. In a second scenario it is attempted to follow the line of constant *cut off* in order to investigate the dependence of the resonance parameters of the Higgs boson on the strength of the quartic coupling. Both strategies belong to distinct physical situations, i.e distinct renormalized quantities. The qualitative picture of the phase transition is shown in figure 2.1. The magnetization gets larger with increasing distance from the critical κ

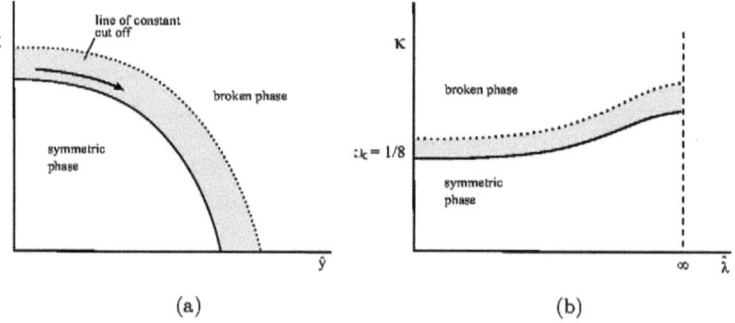

(a) (b)

Figure 2.1: The figure shows the qualitative structure of the phase diagram. The solid line represents the line of critical κ which separates the symmetric phase from the broken phase. Above the critical κ line the mass spectrum does not reflect the symmetries of the Lagrangian and contains a non degenerate scalar particles and massive fermions. The left image (a) shows the phase transition line at zero bare quartic coupling and (b) shows the phase transition at vanishing Yukawa coupling.

line into the broken phase. yielding decreasing cut off values Λ. In this work, all physical results in lattice units are at most half the size of the cut off in lattice units which avoids

2.3 Simulation strategy

unacceptably large cut off effects [48]. Physical results are therefore obtained within a narrow band above the phase transition line. The region is shown in figure 2.1 between the solid phase transition line and the dotted line. Performing simulations at constant values of the cut off and keeping, e.g. the top quark mass fixed in physical units means a fine tuning of the bare parameters of the theory which in practice turns out to be a rather demanding task.

A major part of this work is devoted to the extraction of mass eigenvalues and therefore some details on the determination of masses in lattice quantum field theory and the connection to particle masses from the continuum Minkowski space propagator is given below. The physical mass (m_{phys}) of a particle is given by the eigenvalue of the squared four momentum operator which is an invariant of the Poincaré algebra. The mass can be extracted from the pole of the two point function (propagator) in momentum space

$$G_M^{-1}\left(p^2, m_0^2, \lambda_0, y_0; \tilde{\Lambda}\right) \stackrel{!}{=} 0.$$

$\tilde{\Lambda}$ is a regulator of the theory which is in general not the lattice regulator. In the following the subscript M will indicate the quantity in Minkowski space where the inner product of four vectors are taken with the metric diag $(+1, -1, -1, -1)$.

The Källen-Lehmann representation for the two-point function gives an explicit relation for the propagator in the interacting theory (see e.g. [12]). For a scalar field it is

$$G'_M(x-y) = \left\langle 0 \left| T\left\{\varphi(x)\varphi^\dagger(y)\right\} \right| 0 \right\rangle$$
$$= \int_0^\infty \mathrm{d}s^2 \rho(s^2) \Delta\left(x-y; s^2\right).$$
$$\Delta\left(x-y; s^2\right) := \int \frac{\mathrm{d}^4 k}{(2\pi)^4} e^{-ik(x-y)} \frac{i}{k^2 - s^2 + i\epsilon}.$$

Δ denotes the free scalar propagator of a scalar field with mass s. G'_M is the propagator of the fully interacting theory in Minkowski space time. The Källen-Lehmann relation shows that the propagator in the interacting theory is given by a superposition of free propagators. It can be shown that the spectral weight ρ is positive and real valued. The Fourier transform of the interacting propagator is denoted by $\tilde{G}'_M$ and is given by

$$\tilde{G}'_M(p^2) := \int_{M^2}^\infty \mathrm{d}s^2 \rho(s^2) \frac{i}{p^2 - s^2 + i\epsilon}.$$

M^2 is the smallest eigenvalue of the squared momentum operator P^2 of one particle states excluding the vacuum. In case of a stable particle the one particle states with the energy momentum relation

$$E(\vec{p}) = \sqrt{\vec{p}^2 + m_{phys}^2}$$

can be separated from the spectral weight. The propagator of the interacting theory can then be written as

$$\tilde{G}'_M(p^2) = \frac{iZ}{p^2 - m_{phys}^2 + i\epsilon} + \int_{M_{th}^2}^{\infty} ds^2 \frac{i\overline{\rho}(s^2)}{p^2 - s^2 + i\epsilon}. \tag{2.12}$$

$\overline{\rho}$ is the contribution of multi-particle states and M_{th}^2 is the threshold value of a continuous spectrum of multi-particle states

$$\overline{\rho}(s^2) = 0, \quad s^2 < M_{th}^2.$$

The propagator in the interacting theory has an isolated pole on the real p^2 axis in the limit $\epsilon \to 0^+$ and a branch cut starting from $p^2 = M_{th}^2$ induced by multi-particle states. In a theory where the propagator of an unstable scalar particle is considered, there is no more isolated pole below the branch cut. The branch cut induces new Riemann sheets and the pole of the propagator moves to the second Riemann sheet. In order to locate the pole, the interacting propagator can be continued to complex momenta and one can define $G'_\mathbb{C}(z)$ which is analytic in the complex plane

$$G'_\mathbb{C}(z) := \int_{M^2}^{\infty} ds^2 \rho(s^2) \frac{i}{z - s^2}.$$

The Sokhotsky-Weierstrass theorem yields

$$\lim_{\epsilon \to 0^+} \int_a^b dx \frac{f(x)}{x \pm i\epsilon} = \mathcal{P} \int_a^b dx \frac{f(x)}{x} \mp i\pi f(0)$$

$$\Rightarrow \lim_{\epsilon \to 0^+} \int_{-\infty}^{\infty} dx \frac{\delta(x - x_0)}{x \pm i\epsilon} = \lim_{\epsilon \to 0^+} \frac{1}{x_0 \pm i\epsilon}$$

$$= \underbrace{\mathcal{P} \int_{-\infty}^{\infty} dx \frac{\delta(x - x_0)}{x}}_{0} \mp i\pi \delta(x_0)$$

$$\Rightarrow \lim_{\epsilon \to 0^+} \frac{1}{x \pm i\epsilon} = \mp i\pi \delta(x). \tag{2.13}$$

2.3 Simulation strategy

Using (2.13) the spectral weight is given by the discontinuity across the branch cut

$$\lim_{\epsilon \to 0^+} G'_C(p^2 + i\epsilon) = \int_0^\infty ds^2 \; i\rho(s^2) \lim_{\epsilon \to 0^+} \frac{1}{p^2 - s^2 + i\epsilon}$$

$$= +\pi \, \rho(p^2)$$

$$\lim_{\epsilon \to 0^+} G'_C(p^2 - i\epsilon) = -\pi \, \rho(p^2)$$

$$\Rightarrow \lim_{\epsilon \to 0^+} \left\{ G'_C(p^2 - i\epsilon) - G'_C(p^2 + i\epsilon) \right\} = 2\pi \, \rho(p^2).$$

The pole of G'_C is given by a zero of the inverse propagator

$$G'^{-1}_C = \frac{G'^*_C}{|G'_C|}.$$

The discontinuity of G'^{-1}_C is related to another function $\sigma(x)$ which is related to the spectral weight by

$$\lim_{\epsilon \to 0^+} G'^{-1}_C(p^2 + i\epsilon) \, |G'_C(p^2 + i\epsilon)| = \lim_{\epsilon \to 0^+} G'^*_C(p^2 + i\epsilon) = \pi \, \rho(p^2),$$

$$\lim_{\epsilon \to 0^+} G'^{-1}_C(p^2 - i\epsilon) \, |G'_C(p^2 - i\epsilon)| = -\pi \, \rho(p^2),$$

$$\lim_{\epsilon \to 0^+} \left\{ G'^{-1}_C(p^2 + i\epsilon) - G'^{-1}_C(p^2 - i\epsilon) \right\} = 2\pi \, \sigma(p^2),$$

$$\sigma(p^2) = \frac{\rho(p^2)}{|G'_C(p^2)|}.$$

Finally the analytic continuation of G'^{-1}_C downwards to the second Riemann sheet is defined by

$$G'^{-1}_{II}(p^2 - i\epsilon) := G'^{-1}_C(p^2 + i\epsilon)$$

$$\Rightarrow \lim_{\epsilon \to 0^+} G'^{-1}_{II}(p^2 - i\epsilon) = \lim_{\epsilon \to 0^+} G'^{-1}_C(p^2 - i\epsilon) + 2\pi \, \sigma(p^2).$$

G'^{-1}_{II} has no discontinuity across the real axis. In order to write down an explicit expression one assumes that the parameters of the theory are chosen such that the pole in the second Riemann sheet lies close to the real axis. In view of equation (2.12) G'^{-1}_{II} can be approximated by

$$G'^{-1}_{II}(p^2) \approx \frac{p^2 - m^2_{phys}}{iZ} + 2\pi \, \sigma(p^2)$$

$$\frac{\Gamma}{Z} := 2\pi\sigma(p^2) \geq 0, \qquad 0 < Z \leq 1$$

$$\Rightarrow G'^{-1}_{II}(p^2) \approx \frac{p^2 - m^2_{phys} + i\Gamma}{iZ}. \tag{2.14}$$

The above equation shows that a pole of the propagator in the second Riemann sheet is associated with a non-zero width that is proportional to Γ.

In the framework of Euclidean Monte Carlo simulations, one usually computes the two point correlator which is intimately connected to the time dependence of the Minkowski propagator. The final part of this chapter shows the relation between the correlator and the time dependent propagator.

The time dependence of the analytically continued propagator is given by

$$G'_{\mathbb{C}}(t) = \int \frac{\mathrm{d}p_0}{2\pi} e^{-ip_0 t} G'_{\mathbb{C}}(p^2) \bigg|_{\vec{p}=0}.$$

Using the result (2.14)

$$G'_{\mathbb{C}}(t) = \int_{-\infty}^{\infty} \frac{\mathrm{d}p_0}{2\pi} e^{-ip_0 t} \left\{ \frac{iZ}{p_0^2 - m_{phys}^2 + i\Gamma} + \overline{G}'_{\mathbb{C}}(p_0^2) \right\}.$$

$\overline{G}'_{\mathbb{C}}$ is a smooth function of p^2. The integral can be solved with the help of the residue theorem which yields the sum of the singularities of the integrand within a closed contour. There are two poles in the first term

$$p_0 = \sqrt{m_{phys}^2 - i\Gamma}, \qquad p_0 = -\sqrt{m_{phys}^2 - i\Gamma}.$$

The contour is defined along the real axis and is closed in the lower half of the complex p_0 plane. Figure 2.2 shows a sketch of the contour integral. After some calculations, which are given in the appendix A, the time dependence of the propagator is

$$G'_{\mathbb{C}}(t) = \frac{-Z}{2\sqrt{m_{phys}^2 - i\Gamma}} e^{-i\sqrt{m_{phys}^2 - i\Gamma}\, t} + \ldots$$

$$\Rightarrow G'_{\mathbb{C}}(t) \approx \frac{-Z}{2m_{phys}} e^{-im_{phys} t}\, e^{-\frac{1}{2}\frac{\Gamma}{m_{phys}} t} + \ldots$$

The time dependence is given by an oscillatory factor, which directly contains the physical mass of the particle and an exponential damping factor which is related to the width of the particle. One has to keep in mind that the above result does not refer to the physical propagator. It is the analytically continued propagator and is defined in the complex p_0 plane. The physical propagator of a theory with an hermitian Hamiltonian has real eigenvalues and thus, is only defined for real values of p_0. The Euclidean time correlation function, which is obtained from Monte Carlo simulations are of course physical and as

2.3 Simulation strategy

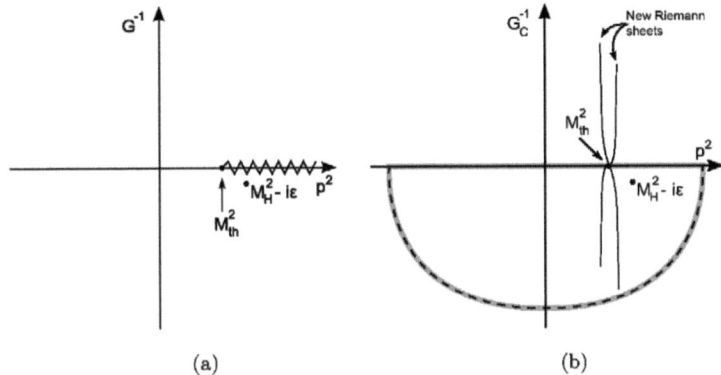

Figure 2.2: The figure shows a sketch of the complex p^2 plane and the contour (blue) along which the integral of p_0 is evaluated. Only the pole in the lower half will contribute to the integral.

such it necessarily involves real values of the energy in the argument of the exponential. The latter statement can easily be derived by using the Hamiltonian as the generator for time evolution and a complete set of eigenstates.

The physical part of the Euclidean Higgs boson propagator can be computed from Monte Carlo data

$$\tilde{G}_H(\hat{p}^2) := \left\langle \tilde{H}_{-\hat{p}} \tilde{H}_{\hat{p}} \right\rangle$$

$$\tilde{H}_p := a^4 \sum_x e^{-i\hat{p}x} H_x.$$

The hat above the momentum variable indicates the discrete lattice momentum $\hat{p} \in \Gamma_{L,T}$

$$\Gamma_{L,T} := \left\{ p \in \mathbb{R}^4 | p_0 = \frac{2\pi}{T} n_0, \quad p_i = \frac{2\pi}{L} n_i, \right.$$

$$\left. n_0 \in \mathbb{Z} : 0 \le n_0 < T, \quad n_i \in \mathbb{Z} : 0 \le n_i < L \right\}$$

Similarly, the Goldstone propagator is given by

$$\tilde{G}_G(\hat{p}^2) := \left\langle \tilde{\mathcal{G}}_{-\hat{p}} \tilde{\mathcal{G}}_{\hat{p}} \right\rangle$$

A fit of the Monte Carlo data to the analytic propagator G'_C for positive momenta, allows to determine the physical pole at some negative or complex momentum. The field renormalization factor for the Goldstone boson is then defined by

$$Z_G^{-1}(-\mu^2) := \left(1 - \frac{\partial}{\partial p^2} \Sigma_G(p^2) \bigg|_{p^2 = -\mu^2} \right)$$

where Σ_G is the self energy contribution of the Goldstone boson, which will be given in Chapter 3.

In the case of unstable or massless particles, it will not be possible to define the mass at the pole of the *physical* propagator. Instead, the renormalization conditions

$$\Re\left\{G_H^{-1}(p^2)\right\}\Big|_{p^2=-M_H^2} = 0, \qquad \Re\left\{G_G^{-1}(p^2)\right\}\Big|_{p^2=-M_G^2} = 0, \qquad (2.15)$$

will be utilized in order to define the renormalized masses. The above definition is consistent with previous work performed within this model [25] and [48].

Finally the fermion time slice correlator will be discussed. As mentioned before, the fermionic degrees of freedom are integrated and are represented by the determinant of the fermion matrix. Correspondingly, the time slice correlator of fermions is not directly accessible. The fermion propagator and thus the fermion correlator can be constructed from the matrix elements of $\mathcal{M}^{-1}$. The fermion mass can be extracted from either the left or the right handed components of the spinor. In the following arguments only the left handed components are discussed, but it is straight forward to apply the arguments for the right handed components. The left handed correlator $C_f(t_0 - t_1)$ is given by

$$C_f(t_0 - t_1) := \left\langle \text{Tr}\left\{ \left(\hat{P}_L \Psi\right)_{(t_0,\vec{p}=0)} \left(\bar{\Psi} P_L\right)_{(t_1,\vec{p}=0)} \right\} \right\rangle.$$

The $\vec{p} = 0$ component is obtained by performing a sum over all space time points. In fact, this corresponds to a Fourier transformation

$$\left\langle \text{Tr}\left\{ \left(\hat{P}_L \Psi_i\right)_{(t_0,\vec{p}=0)} \left(\bar{\Psi}_i P_L\right)_{(t_1,\vec{p}=0)} \right\} \right\rangle = \frac{1}{V^2} \sum_{\alpha=1}^{4} \sum_{\vec{x},\vec{y}} \hat{P}_L^{t_0,\vec{x},\alpha,i;\, t_0',\vec{x}',\alpha',i}$$

$$\left\langle \Psi_{t_0',\vec{x}',\alpha',i}\, \bar{\Psi}_{t_1',\vec{y}',\beta',i} \right\rangle P_L^{t_1',\vec{y}',\beta',i;\, t_1,\vec{y},\alpha,i}$$

$$= \frac{1}{V^2} \sum_{\alpha=1}^{4} \sum_{\vec{x},\vec{y}} \hat{P}_L^{t_0,\vec{x},\alpha,i;\, t_0',\vec{x}',\alpha',i}$$

$$\mathcal{M}^{-1}_{t_0',\vec{x}',\alpha',i;\, t_1',\vec{y}',\beta',i} P_L^{t_1',\vec{y}',\beta',i;\, t_1,\vec{y},\alpha,i}$$

The index i at the fermion fields Ψ denotes either the top or the bottom quark. Within this work, it will not play a role, as a degenerate quark doublet is assumed. The matrix elements of the inverse fermion matrix is then obtained by a conjugate gradient (CG) algorithm. The starting- and the solution vector of the CG algorithm are give according

2.3 Simulation strategy

to

$$\mathcal{M}^\dagger \mathcal{M} \cdot u = \mathcal{M}^\dagger P_L \cdot v$$
$$\Rightarrow u = \mathcal{M}^{-1} P_L \cdot v$$
$$\Rightarrow \hat{P}_L \cdot u = \hat{P}_L \mathcal{M}^{-1} P_L \cdot v$$

The vector v is chosen such that on time slice t', spinor index α and all $\vec{b} \in \mathbb{R}^3$, v is equal to 1, i.e.

$$\forall \vec{b} \in \mathbb{R}^3 : \quad v\left(\{t', \vec{b}\}, \alpha\right) \equiv 1$$

With the above choice of v, a sum over the space time is automatically performed. The CG algorithm returns the vector u, after the multiplication of the chiral projector $\hat{P}_L$ from the left and performing a sum over all space time points one finally obtains the fermion correlator.

The top and bottom quarks in this model are stable particles and hence the time slice correlator will be utilized in order to extract the fermion masses. With the correlator at hand, one can either fit an exponential function and extract the fermion mass from the argument of the exponential or one can compute the effective masses given by

$$\log C^{eff}(t) := \log \frac{C(t+1)}{C(t)} = \log \left\{ e^{-m_f(t+1)} e^{m_f t} \right\} = -m_f.$$

Due to the hyper cubic symmetry and periodic boundary condition of the fields, the measurements of correlators on the lattice follow rather a hyperbolic cosine function than an exponential

$$C(\Delta t) := \cosh\left(m\left(\Delta t - \frac{T}{2}\right)\right).$$

T is the lattice extent in temporal direction. Hence the effective masses C^{eff} are ratios of cosine hyperbolicus

$$C^{eff}(\Delta t) = \frac{C(\Delta t + 1)}{C(\Delta t)}$$
$$= \frac{\cosh\left(m\left(\Delta t + 1 - \frac{T}{2}\right)\right)}{\cosh\left(m\left(\Delta t - \frac{T}{2}\right)\right)}.$$

Using the identity

$$\cosh(x+y) = \cosh(x)\cosh(y) + \sinh(x)\sinh(y)$$

C^{eff} can be written as

$$C^{eff}(\Delta t) = \frac{1}{\cosh\left(m\left(\Delta t - \frac{T}{2}\right)\right)} \left\{ \cosh(m)\cosh\left(m\left(\Delta t - \frac{T}{2}\right)\right) + \sinh(m)\sinh\left(m\left(\Delta t - \frac{T}{2}\right)\right) \right\}$$

$$= \cosh(m) + \sinh(m)\tanh\left(m\left(\Delta t - \frac{T}{2}\right)\right).$$

For large arguments $|m\left(\Delta t - \frac{T}{2}\right)| \gg 1$ the hyperbolic tangent can be approximated

$$\tanh(x) = \frac{\sinh(x)}{\cosh(x)}$$
$$= \frac{e^x - e^{-x}}{e^x + e^{-x}}, \qquad \text{assuming } x \gg 1$$
$$= \frac{e^x(1 - e^{-2x})}{e^x(1 + e^{-2x})} \to 1.$$

In the case $x \ll 1$ an anlogous calculation yield $\tanh(x) \to -1$

$$\Rightarrow C^{eff} = \begin{cases} \cosh(m) - \sinh(m) = e^{-m}, & m \ll 1 \\ \cosh(m) + \sinh(m) = e^{+m}, & m \gg 1 \end{cases}$$

The above calculation shows that the effective masses computed on the lattice indeed represents an exponential even though the correlators itself do not. It is useful to choose $|\Delta t|$ away from $\frac{T}{2}$ in order to justify the approximation $\tanh(\Delta t - \frac{T}{2}) \to \pm 1$. Figure 2.3 shows the fermion correlator and the logarithm of the effective masses for a selected run at $y_0 = 3.12305$ and infinite bare quartic coupling. The data was computed on a $16^3 \times 32$ lattice. The given fermion masses presented in this work are computed with the help of the effective masses.

2.3 Simulation strategy

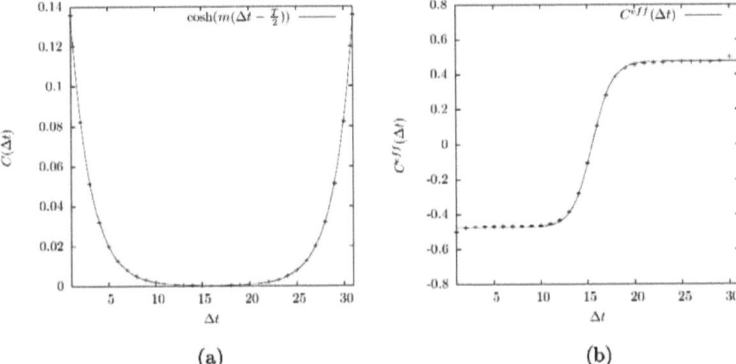

Figure 2.3: The figure shows the fermion correlator (a) and the logarithm of the effective masses (b) for a selected run with bare parameters $y_0 = 3.12305$ and infinite bare quartic coupling. The data was computed on a $16^3 \times 32$ lattice.

3 Analytic properties and perturbative calculations

Section 3.1 of this chapter summarizes perturbation theory in the Euclidean continuum and some details on the renormalization procedure are presented. Finally the renormalized expressions for the Higgs boson propagator as well as the Goldstone boson propagator are given. The next section deals with lattice perturbation theory and focuses especially on the perturbative expansion of the fermion-Higgs coupling. The expansion takes care about the modifies chiral lattice projectors and the Neuberger overlap operator. Finally the results are compared with results obtained from Monte Carlo simulations.

3.1 Perturbative expansion in the continuum

This section deals with perturbation theory in the Euclidean continuum. The perturbative expansion in the renormalized couplings of the theory is a powerful tool as long as those couplings are small enough. Perturbation theory can provide analytic expressions for observables such as the propagator. It is the aim of this chapter to investigate the perturbative predictions in the Higgs-Yukawa model in order to compare them with those obtained from numerical simulation. In the case of small couplings, the analytic expression suggested by perturbation theory will be utilized to determine the pole of the propagator after fitting its parameters to the data obtained by the Monte Carlo simulations. It turns out that the functional form of the scalar propagators can describe the numerical data even for large bare couplings very well.

The starting point is the Lagrangian given in (2.1) and is repeated here

$$L_E^{HY} = \frac{1}{2}(\partial_\mu \varphi)^\dagger \cdot (\partial^\mu \varphi) + \frac{1}{2}m^2 \varphi^\dagger \cdot \varphi + \lambda \left(\varphi^\dagger \cdot \varphi\right)^2$$

$$+ \bar{t} \slashed{D} t + \bar{b} \slashed{D} b + y_b \begin{pmatrix} \bar{t} \\ \bar{b} \end{pmatrix}_L^T \cdot \varphi \, b_R + y_t \begin{pmatrix} \bar{t} \\ \bar{b} \end{pmatrix}_L^T \cdot \tilde{\varphi} \, t_R + h.c..$$

The footing of perturbation theory is the Lehmann-Symanzik-Zimmermann (LSZ) reduction formula which relates transition probabilities to Green functions of the theory. The derivation is given in many text books about quantum field theory e.g. [42]

$$\langle k_1 s_1; \ldots; k_m s_m; out | p_1 s_1'; \ldots; p_n s_n'; in \rangle =$$

$$(2\pi)^4 \delta^4 \left(\sum_{i=1}^m k_i - \sum_{j=1}^n p_j\right) \left(\sqrt{Z}\right)^{n+m} \prod_{i=1}^m \bar{u}_{\alpha_i}(k_i, s_i)$$

$$\times \tilde{G}_{amp}(k_1, \ldots, k_m; -p_1, \ldots, -p_n)_{\alpha_m, \ldots, \alpha_1, \beta_1, \ldots, \beta_n} \prod_{j=1}^n u_{\beta_j}\left(p_j; s_j'\right).$$

The left hand side denotes the matrix element of asymptotic 'in' and 'out' fields in momentum space. s and s' are the spin eigenvalues of the field operators and $\tilde{G}_{amp}$ is the Fourier transform of the amputated connected Green function which can be expanded with the help of Feynman diagrams. u denotes the usual spinors in the case of fermions; in the case of scalar fields they are identical to one. The Green functions of the theory are given by the path integral in equation (2.3)

$$\langle \Omega | T \{ \mathcal{O}(x_a) \mathcal{O}(x_a) \cdots \} | \Omega \rangle = \mathcal{N} \int D\varphi D\{t, \bar{t}, b, \bar{b}\} \, e^{-S(\varphi, t, \bar{t}, b, \bar{b})} \mathcal{O}(x_a) \mathcal{O}(x_a) \cdots$$

The expansion in terms of Feynman diagrams makes use of the fact that the action can be split up in a free Gaussian part and an interaction term

$$S\left(\varphi, t, \bar{t}, b, \bar{b}\right) = S_0^\varphi(\varphi) + S_0^f\left(t, \bar{t}, b, \bar{b}\right) + S_I\left(\varphi, t, \bar{t}, b, \bar{b}\right)$$

$$=: S_0^{\varphi, t, b} + S_I^{\varphi, t, b}.$$

The interaction is treated as a perturbation such that the exponential series in S_I can be truncated after some finite terms. The resulting integral is Gaussian with some polynomial factor and in principle such an integral can be calculated exactly

$$\langle \Omega | T \{ \mathcal{O}(x_a) \mathcal{O}(x_a) \cdots \} | \Omega \rangle = \mathcal{N} \int D\varphi D\{t, \bar{t}, b, \bar{b}\} \, e^{-S_0^{\varphi, t, b}(\varphi, t, \bar{t}, b, \bar{b})}$$

$$\prod_{n=0}^N \frac{1}{n!} \left\{-S_I\left(\varphi, t, \bar{t}, b, \bar{b}\right)\right\}^n \mathcal{O}(x_a) \mathcal{O}(x_a) \cdots \quad (3.1)$$

3.1 Perturbative expansion in the continuum

The aim of this chapter is to derive a perturbative expression for the Goldstone and the Higgs boson propagator in Euclidean field theory. All calculations are therefore restricted to the broken phase where the scalar expectation value is non zero. Furthermore it is assumed, that the scalar *vev* takes the formula

$$\langle \varphi \rangle = \begin{pmatrix} 0 \\ \frac{v}{\sqrt{2}} \end{pmatrix}.$$

The Higgs boson and the Goldstone boson fields are then defined by

$$\varphi = \begin{pmatrix} \mathcal{G}_1 + i\mathcal{G}_2 \\ v + H + i\mathcal{G}_3 \end{pmatrix} \tag{3.2}$$

where the physical fields $H, \mathcal{G}_i$ have vanishing vacuum expectation value.

The quality i.e. the convergence property, of a perturbative series strongly depends whether one is able to expand at the right minimum. In the broken phase the minimum of the action is shifted away from the origin and it is necessary to perform the perturbative expansion with respect to the shifted field h in order to obtain reliable results with the leading order terms. The Lagrangian rephrased in the physical fields $H, \mathcal{G}_i$ is given by

$$\begin{aligned}\mathcal{L} = \bar{t}\slashed{D}t + yv\,\bar{t}t + \bar{b}\slashed{D}b + yv\,\bar{b}b + \frac{1}{2}(\nabla H)^2 + \frac{1}{2}\left(8\lambda v^2\right)H^2 + \frac{1}{2}\nabla \mathcal{G}^T \nabla \mathcal{G} \\ + \lambda H^4 + \lambda \left(\mathcal{G}^T\mathcal{G}\right)^2 + yH\left(\bar{t}t\right) + yH\left(\bar{b}b\right) + 2\lambda \left(\mathcal{G}^T\mathcal{G}\right)H^2 \\ + 4\lambda v\, \mathcal{G}^T\mathcal{G}H + 4v\lambda\, H^3 + y\Big\{G_1\left(\bar{t}_L b_R - \bar{b}_L t_R\right) \\ + iG_2\left(\bar{b}_L t_R + \bar{t}_L b_R\right) + iG_3\left(\bar{b}_L b_R - \bar{t}_L t_R\right) + h.c.\Big\}.\end{aligned} \tag{3.3}$$

The minimum of the potential at tree level is given by

$$m^2 + 4\lambda v^2 = 0.$$

The broken phase thus contains a massive scalar particle and massive fermions. The bare masses are

$$m_\varphi^2 = 8\lambda v^2 = -2m^2, \quad m^2 < 0$$

$$m_f = y\, v.$$

It is well known that the full two point function can be expressed in a geometric series. The bare propagator can then be written in terms of the one particle irreducible part Σ

$$G^{-1}\left(p^2, m^2, \ldots; \Lambda\right) = p^2 + m^2 - \Sigma\left(p^2, m^2, \ldots; \Lambda\right).$$

In the following the Goldstone and the Higgs boson self energy contributions are computed up to one loop with a finite energy cut off as a regulator for the Feynman integrals.

3.1.1 The Higgs boson propagator

From the Lagrangian (3.3) one infers the interaction terms which contribute to the one particle irreducible Feynman diagrams of the Higgs boson. Table 3.1 shows an overview of the relevant interaction terms and their Wick contractions. The first column identifies the corresponding Feynman graph, which is displayed in figure 3.1. The second column shows the coupling in the Lagrangian (3.3) from which the specific interaction arises while the third column shows the order in the perturbative expansion of the interaction term. The next column shows one of the possible Wick contractions associated to the interaction term. There are in general several Wick contractions to each such interaction term which however, lead to the same analytic expression. Feynman diagrams are symbolic representations of the Wick contractions and those diagrams which lead to identical analytical expressions are topologically equivalent. The last column of the table gives the symmetry factor which denotes the number of topologically equivalent Feynman diagrams.

The expansion of the two point function in equation (3.1) up to the second order gives

$$\langle \Omega | T\{H(x)H(y)\} | \Omega \rangle = \mathcal{N} \int D\varphi D\{t, \bar{t}, b, \bar{b}\} \, e^{-S_0^{\varphi,t,b}(\varphi, t, \bar{t}, b, \bar{b})}$$
$$\left\{ 1 - S_I\left(\varphi, t, \bar{t}, b, \bar{b}\right) + \frac{1}{2} S_I^2\left(\varphi, t, \bar{t}, b, \bar{b}\right) \right\} H(x) H(y).$$

The above relations correspond to the position space two point Green function. As one is interested in the momentum space propagator, a Fourier transform has to be performed. The spatial integrals within the action $S_I = \int d^4x \mathcal{L}_{int}$ and the exponentials arising from the position space representation of the propagators can be performed which lead to δ distributions which in turn makes it easy to perform some of the momentum space integrals. The δ distribution represents the translation invariance or equivalently the total momentum conservation. This procedure is well known and can be found in many text books on quantum field theory e.g. [42]. The momentum space two point function then reads

$$\int d^4x d^4y \, e^{ip_1 x + ip_2 y} \langle \Omega | T\{H(x)H(y)\} | \Omega \rangle = (2\pi)^4 \delta(p_1 + p_2) \tilde{G}^{(2)}(p_1, p_2).$$

3.1 Perturbative expansion in the continuum

Table 3.1: The table below shows the relevant interaction terms in the Lagrangian which contribute to the one particle irreducible diagrams of the Higgs boson propagator. The first column is an identifier for the corresponding diagrams shown in figure 3.1. S_I is a part of the interaction Lagrangian. The middle column shows the order in the expansion of the exponential of the interaction term. The next column displays a Wick contraction and its multiplicity is given in the last column.

	S_I	Perturbative expansion	Wick contraction type	fac.
A	λH^4	$-\lambda \left(H^4\right)_{x_1}$	$\overline{H_x \, (\overline{HHHH})_{x_1} \, H_y}$	12
B	$2\lambda \left(\mathcal{G}^T\mathcal{G}\right) H^2$	$-2\lambda \left(\mathcal{G}\mathcal{G}H^2\right)_{x_1}$	$\overline{H_x \left(\overline{\mathcal{G}^T\mathcal{G}H^2}\right)_{x_1} H_y}$	2
C	$4\lambda v H^3$	$\frac{(4\lambda v)^2}{2} \left(H^3\right)_{x_1} \left(H^3\right)_{x_2}$	$\overline{H_x \, (\overline{HHH})_{x_1} \, (\overline{HHH})_{x_2} \, H_y}$	36
D	$4\lambda v \left(\mathcal{G}^T\mathcal{G}\right) H$	$\frac{(4\lambda v)^2}{2} (\mathcal{G}\mathcal{G}H)_{x_1} (\mathcal{G}\mathcal{G}H)_{x_2}$	$\overline{H_x (\overline{\mathcal{G}^T\mathcal{G}H})_{x_1} (\overline{\mathcal{G}^T\mathcal{G}H})_{x_2} H_y}$	4
E	$y \left(H \bar{t} t\right)$	$\frac{y^2}{2} (H\bar{t}t)_{x_1} (H\bar{t}t)_{x_2}$	$\overline{H_x \, (\overline{H\bar{t}t})_{x_1} \, (\overline{H\bar{t}t})_{x_2} \, H_y}$	2
F	$y \left(H \bar{b} b\right)$	$\frac{y^2}{2} (H\bar{b}b)_{x_1} (H\bar{b}b)_{x_2}$	$\overline{H_x (\overline{H\bar{b}b})_{x_1} (\overline{H\bar{b}b})_{x_2} H_y}$	2

The Wick contraction of the scalar fields are

$$\overline{H(x)\, H(y)} = \Delta_H(x-y)$$
$$= \int \frac{d^4 q}{(2\pi)^4} \, e^{-iq(x-y)} \frac{1}{q^2 + m_\varphi^2}$$

$$\overline{\mathcal{G}^T(x)\, \mathcal{G}(y)} = \sum_{i=1}^{3} \overline{\mathcal{G}_i^T(x)\, \mathcal{G}_i(y)}$$
$$= 3 \int \frac{d^4 q}{(2\pi)^4} \, e^{-iq(x-y)} \frac{1}{q^2 + m_G^2}.$$

Though the bare Goldstone boson mass as well as the renormalized Goldstone boson mass vanishes an explicit mass term is kept which can be set to zero. It is the aim to use the perturbative form of the propagator as a fit function for the numerical Monte Carlo data. The Goldstone bosons acquire a mass in finite volume and thus it will be valuable to keep the explicit mass term in the following derivation.

The leading terms in the propagator are given by

$$\int d^4x d^4y \, e^{ip_1 x + ip_2 y} \langle \Omega | T \{H(x) H(y)\} | \Omega \rangle = \int d^4x d^4y \, e^{ip_1 x + ip_2 y}$$
$$\left\{ \overline{H(x) H(y)} - 12\lambda \int d^4x_1 \, \overline{H_x \, (\overline{HHHH})_{x_1} \, H_y} \ldots \right\}$$

$$= (2\pi)^4 \, \delta(p_1 + p_2) \left\{ \tilde{\Delta}(p_1) - 12\lambda \, \tilde{\Delta}(p_1) \underbrace{\left[\int \frac{d^4 q}{(2\pi)^4} \frac{1}{q^2 + m_\varphi^2} \right]}_{=: D(m_\varphi^2; \Lambda)} \tilde{\Delta}(p_2) + \ldots \right\}$$

The Wick contractions in 3.1-A and 3.1-B describes tadpole diagrams and will be abbreviated by $D(m^2; \Lambda)$ where Λ is a regulator of the integral. The scalar loops which arise from the contractions in 3.1-(C,D) gives the first contribution to the two point function which depends on the external momenta. The momentum space representation is given by

$$I_1(p^2, m^2, \Lambda) := \int \frac{d^4 q}{(2\pi)^4} \frac{1}{q^2 + m^2} \frac{1}{(p+q)^2 + m^2}.$$

The contraction of the fermion fields yields

$$\overline{t_\alpha \, \bar{t}_\beta} = -\overline{\bar{t}_\beta \, t_\alpha} = \int \frac{d^4 q}{(2\pi)^4} \, e^{-iq(x-y)} \left(\frac{1}{\slashed{q} + m_f} \right)_{\alpha\beta}.$$

The contraction related to the Higgs-Yukawa coupling as shown in table 3.1E and 3.1F gives a trace of the fermion matrices

$$\overline{(\bar{t}_\alpha \, t_\alpha)_{x_1} \cdot (\bar{t}_\beta \, t_\beta)_{x_2}} = \int \frac{d^4 q \, d^4 k}{(2\pi)^8} \left(\frac{-e^{-iq(x_2-x_1)}}{\slashed{q} + m_f} \right)_{\beta\alpha} \left(\frac{e^{-ik(x_1-x_2)}}{\slashed{k} + m_f} \right)_{\alpha\beta}.$$

The contribution of the fermion loop to the momentum space, one particle irreducible, two point function yields

$$J(p^2, m_f, \Lambda) := -\int \frac{d^4 q}{(2\pi)^4} \, \mathrm{Tr} \left[\left(\frac{1}{\slashed{p} + \slashed{q} + m_f} \right) \left(\frac{1}{\slashed{q} + m_f} \right) \right], \quad q^0 < \Lambda$$

Where p is again an external momenta.

Though the tadpole diagrams can be calculated easily, an explicit expression will not be given, as they cancel during the renormalization procedure anyway. The self energy contribution to the Higgs boson up to one loop is

$$\Sigma(p^2, m_H^2, m_G^2, \lambda, y; \Lambda) = -12\lambda \, D(m_\varphi^2) - 12\lambda \, D(m_G^2)$$
$$+ 18 \, (4\lambda v)^2 \, I_1(p^2, m_\varphi^2, \Lambda) + 6 \, (4\lambda v)^2 \, I_1(p^2, m_G^2, \Lambda) + y^2 J(p^2, m_f, \Lambda) \quad (3.4)$$

3.1 Perturbative expansion in the continuum

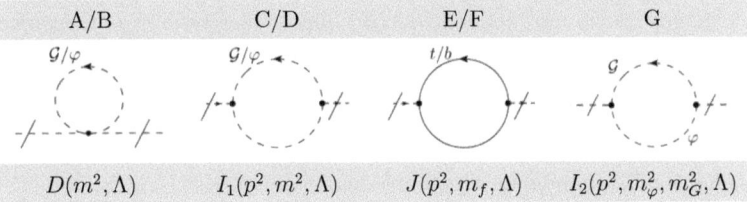

Figure 3.1: The figure shows the different type of one loop integrals contributing to the one particle irreducible two point function of scalar particles. The dashed lines denote a scalar propagator while the solid lines correspond to fermion propagators.

The explicit expressions for the integrals are given below. The bare Goldstone boson mass and in fact the renormalized Goldstone boson mass vanishes due to the Goldstone theorem. Here the mass m_G will be kept explicitly since the final aim is to compare the perturbative results with the numerical studies where due to the finite volume Goldstone bosons are not exactly massless. Furthermore, the explicit appearance of the Goldstone mass will be helpful to check the consistency of the calculation since it provides an infrared regulator.

As was mentioned in Chapter 2, there are some subtleties connected with unstable particles. For the calculations below, it is therefore assumed that the Higgs boson is stable. The bare parameters of the model shall be chosen such that the Higgs boson does not decay into two Goldstone bosons. This also implies that artificial Goldstone boson mass takes a non zero positive value. The renormalization condition for the Higgs boson propagator is then chosen to be

$$G_H^{-1}\left(p^2 = -\mu^2, m_H^2, m_G^2, \lambda, y; \Lambda\right) = 0.$$

μ is an arbitrary mass scale which will be discussed later. The above renormalization condition can be inverted and leads to an expression in terms of the renormalized mass μ^2. It is not necessary to solve the above equation for μ^2, instead it is sufficient to get an expression of μ^2 which is correct up to the considered order in the perturbative expansion.

$$G_H^{-1}\left(p^2 = -\mu^2, m_H^2, m_G^2, \lambda, y; \Lambda\right) = -\mu^2 + m_H^2 - \Sigma\left(-\mu^2, m_H^2, m_G^2, \lambda, y; \Lambda\right)$$
$$\Rightarrow m_H^2 = \mu^2 + \Sigma\left(-\mu^2, m_H^2, m_G^2, \lambda, y; \Lambda\right).$$

Though the self energy term contains the bare Higgs boson mass m_H, it is perfectly fine to replace them by the renormalized mass μ. The error induced by this substitution is of order λ^4 while the above relation is of order λ^2 in the expansion of the interaction Lagrangian. Taking the tree level relations between the bare quartic and the Yukawa couplings which are denoted by λ_R, y_R, the renormalized Higgs boson propagator is then given by

$$Z_H \left\{ G_H^R \left(p^2, M_H^2, M_G^2, \lambda_R, y_R; \Lambda \right) \right\}^{-1} = p^2 + \mu^2 -$$
$$\left\{ \Sigma \left(p^2, \mu^2, M_G^2, \lambda_R, y_R; \Lambda \right) - \Sigma \left(-\mu^2, \mu^2, M_G^2, \lambda_R, y_R; \Lambda \right) \right\}.$$

Z_H is the Higgs boson field renormalization factor. The renormalized propagator has the form

$$G_H^R \left(p^2, M_H^2, \ldots \right) = \frac{1}{p^2 + M_H^2 - \Delta\Sigma_H \left(p^2, M_H^2, \ldots \right)}$$
$$= \langle H_R(-p) H_R(p) \rangle$$

In order to sustain the above form, the renormalized fields have to be introduced. They are given by

$$H_R := \frac{1}{\sqrt{Z_H}} H.$$

It can be shown that the field renormalization factor Z_H only depends on p^s and due to O(4) invariance (respectively Poincaré invariance in Minkowski space), the Z factor must be a constant [42]. Λ still denotes the implicit cut off of the theory. Here the dependence of the cut off Λ will be kept explicit to indicate that the theory has only a Gaussian fix point, i.e. the theory is trivial. One can see that the renormalized propagator contains the difference of self energy terms and as will be shown below, all divergences up to the considered order in perturbation theory are cancelled within the difference such that the renormalized propagator is finite. Here the renormalization point will be chosen at the physical Higgs boson mass $\mu^2 = \left(m_{phys}^H \right)^2 =: M_H^2$ which is also known as the on-shell renormalization scheme.

In the following explicit expressions for the integrals are presented. A detailed calculation is given in the appendix B. The scalar one loop integral with a single bosonic mass

3.1 Perturbative expansion in the continuum

term yields

$$I_1(p^2, m^2, \Lambda) = \int \frac{d^4q}{(2\pi)^4} \frac{1}{q^2 + m^2} \frac{1}{(p+q)^2 + m^2}, \quad q^0 < \Lambda$$

$$= \frac{1}{(4\pi)^2} \left\{ 1 + \ln\left(\frac{\Lambda^2}{m^2}\right) \right.$$

$$\left. - \sqrt{1 + \frac{4m^2}{p^2}} \ln\left(\frac{1 + \sqrt{1 + \frac{4m^2}{p^2}} - i\epsilon \, \text{Sgn}(p^2)}{-1 + \sqrt{1 + \frac{4m^2}{p^2}} - i\epsilon \, \text{Sgn}(p^2)}\right) \right\}.$$

The $i\epsilon$- terms are kept explicitly although in Euclidean space the limit $\epsilon \to 0$ could safely be performed. Finally the propagator has to be analytically continued to Minkowski space time where these $i\epsilon$ terms will be helpful in order to choose the right contour integrals around the singularities.

The fermionic integral evaluates to

$$J(p^2, m_f, \Lambda) = \int \frac{d^4q}{(2\pi)^4} \text{Tr} \left[\frac{1}{(\not{p} + \not{q} + m_f)} \frac{1}{(\not{q} + m_f)} \right], \quad q^0 < \Lambda$$

$$= \frac{1}{4\pi^2} \left\{ \Lambda^2 + 3m_f^2 \log\left(\frac{\Lambda^2}{p^2}\right) - m_f^2 \right.$$

$$\left. - \frac{1}{2}p^2 \left(\log\left(\frac{\Lambda^2}{p^2}\right) - 6\Delta I_3\left(p^2, \frac{m_f^2}{p^2}\right) - \frac{1}{3}\right) \right\}, \quad (3.5)$$

$$\Delta I_3(p^2, \frac{m_f^2}{p^2}) = \frac{m_f^2}{p^2} - \frac{5}{3} + \frac{11}{12} \log\left(\frac{m_f^2}{p^2}\right)$$

$$+ \frac{1}{8}\left(7 - \frac{4m_f^2}{p^2}\right)\sqrt{1 + \frac{4m_f^2}{p^2}} \log\left(\frac{\sqrt{1 + \frac{4m_f^2}{p^2}} + 1}{\sqrt{1 + \frac{4m_f^2}{p^2}} - 1}\right).$$

As mentioned before the tadpole diagrams do not contribute to the renormalized propagator and just the definition is given

$$D(m^2) = \int \frac{d^4q}{(2\pi)^4} \frac{1}{q^2 + m^2}.$$

The renormalized Higgs boson propagator depends on the difference of the one particle irreducible parts

$$\Delta\Sigma\left(-\mu^2, \mu^2, M_G^2, \lambda_R, y_R; \Lambda\right) =$$

$$\Sigma\left(p^2, \mu^2, M_G^2, \lambda_R, y_R; \Lambda\right) - \Sigma\left(-\mu^2, \mu^2, M_G^2, \lambda_R, y_R; \Lambda\right).$$

Inserting (3.4) yields

$$\Delta\Sigma\left(p^2, M_H^2, M_G^2, \lambda_R, y_R; \Lambda\right) = +18\left(4\lambda_R v\right)^2 \left(I_1(p^2, M_H^2, \Lambda) - I_1(-M_H^2, M_H^2, \Lambda)\right)$$
$$+ 6\left(4\lambda_R v\right)^2 \left(I_1(p^2, M_G^2, \Lambda) - I_1(-M_H^2, M_G^2, \Lambda)\right)$$
$$+ y_R^2 \left(J(p^2, M_f, \Lambda) - J(-M_H^2, M_f, \Lambda)\right).$$

The capital letters $M_{H,G,f}$ denote the physical mass of the scalar Higgs boson and the Goldstone bosons while the subscript f stands for the quarks. The differences in the brackets involving the function I_1 are independent of the regulator Λ. The contribution of the fermion is more subtle as it contributes to the field renormalization factor of the Higgs boson.

The contribution to the Z factor can be seen by decomposing the fermionic contribution to the Higgs boson self energy in its invariant constituents with respect to the Euclidean symmetry

$$\Sigma_H^f(p^2, m_f, y, \Lambda^2) := -y^2 \underbrace{\left(a(p^2)\, p^2 + b(p^2)\, m_f^2\right)}_{J(p^2, m_f, \Lambda)}. \tag{3.6}$$

where a, b are dimensionless scalar functions. Both functions can be expanded around $p^2 = -M_H^2$. From the expression (3.5), it is clear that both scalar functions are divergent in the limit $\Lambda \to \infty$. While divergences in b can be absorbed into the bare mass m_H, the divergences in a cannot. The divergent term in a is logarithmic and $\partial_{p^2} a(p^2)$ is finite. To isolate the divergent part, an expansion of the functions a, b at $p^2 = -M_H^2$ gives

$$a(p^2) = \underbrace{a(p^2 = -M_H^2)}_{:=a_1} + \underbrace{(p^2 + M_H^2)\frac{\partial}{\partial p^2} a(p^2)\bigg|_{p^2=-M_H^2}}_{:=\tilde{a}(p^2)} + \ldots$$

$$b(p^2) = \underbrace{b(p^2 = -M_H^2)}_{:=b_1} + \underbrace{(p^2 + M_H^2)\frac{\partial}{\partial p^2} b(p^2)\bigg|_{p^2=-M_H^2}}_{:=\tilde{b}(p^2)} + \ldots$$

$$\tag{3.7}$$

a_1 and b_1 are divergent as Λ becomes large. In the above case

$$a_1 = -\frac{1}{4\pi^2}\left\{\frac{1}{2}\log\left(\frac{\Lambda^2}{-M_H^2}\right) - 3\Delta I_3(-M_H^2) - \frac{1}{6}\right\}$$
$$b_1 = \frac{1}{4\pi^2}\left\{\frac{\Lambda^2}{m_f^2} + 3\log\left(\frac{\Lambda^2}{-M_H^2}\right) - 1\right\}.$$

3.1 Perturbative expansion in the continuum

The field renormalization factor is defined by the partial derivative of the self energy with respect to the squared momentum. At the considered order in perturbation theory, there is no contribution from the scalar loop integrals. From the above calculation it is clear that the term a_1 contains the fermionic contribution to the Z_H factor as it is proportional to p^2.

$$G_H^{-1}\left(p^2, m_H^2, m_G^2, \lambda, y; \Lambda\right) = p^2 + m_H^2 - \Sigma_H^{H,G}\left(m_H^2, m_G^2, \lambda; \Lambda\right) - \Sigma_H^f\left(m_f, y; \Lambda\right).$$

$\Sigma_H^{H,G}$ denotes the contributions of the Higgs boson self interaction and the interactions to the Goldstone bosons to the self energy of the Higgs boson. The Higgs boson field renormalization factor shall be given by

$$Z_H := 1 - \frac{\partial}{\partial p^2} \Sigma_H^f\left(p^2, m_f, y; \Lambda\right)\bigg|_{p^2 = -M_H^2}$$
$$- \frac{\partial}{\partial p^2} \Sigma_H^{H,G}\left(p^2, m_H^2, m_G^2, \lambda; \Lambda\right)\bigg|_{p^2 = -M_H^2}$$
$$= 1 + y^2 a_1$$
$$\Rightarrow \frac{1}{Z_H} = 1 - y^2 a_1 + \mathcal{O}\left(y^4\right).$$

In the last step the representations given in (3.6) and (3.7) was used. $\tilde{a}$ an $\tilde{b}$ vanish as the expression is evaluated at $p^2 = -M_H^2$.

$$\frac{1}{Z_H} G^{-1}\left(p^2, m_H^2, m_G^2, \lambda, y; \Lambda\right) =$$
$$\left(1 - y^2 a_1\right) \left\{ p^2 + m_H^2 + y^2 \left(p^2 (a_1 + \tilde{a}) + m_f^2 (b_1 + \tilde{b})\right) - \Sigma_H^{H,G}\left(p^2, m_H^2, m_G^2, \lambda; \Lambda\right) \right\}.$$

Neglecting all terms of order y^4 and $y^2 \lambda^2$ such as $y^2 a_1 \Sigma_H^{H,G}$, the above relation reduces to

$$\frac{G^{-1}\left(p^2, m_H^2, m_G^2, \lambda, y; \Lambda\right)}{Z_H} = p^2 + m_H^2 + m_f^2 y^2 (b_1 - a_1)$$
$$+ y^2 \left(\tilde{a}(p^2) p^2 + \tilde{b}(p^2) m_f^2\right) - \Sigma_H^{H,G}\left(p^2, m_H^2, m_G^2, \lambda; \Lambda\right)$$
$$\Rightarrow \Sigma_H^f\left(p^2, m_f, y; \Lambda\right) = -y^2 \left\{ m_f^2 (b_1 - a_1) + \tilde{a}(p^2) p^2 + \tilde{b}(p^2) m_f^2 \right\} + \mathcal{O}\left(y^4\right)$$
$$\Rightarrow \Delta\Sigma_H^f\left(p^2, m_f, y; \Lambda\right) := \Sigma_H^f\left(p^2, m_f, y; \Lambda\right) - \Sigma_H^f\left(-\mu^2, m_f, y; \Lambda\right)$$
$$= -y^2 \left\{ \left(\tilde{a}(p^2) - \tilde{a}(-\mu^2)\right) p^2 \right.$$
$$\left. + \left(\tilde{b}(p^2) - \tilde{b}(-\mu^2)\right) m_f^2 \right\} + \mathcal{O}\left(y^4\right).$$

In the above expression for the self energy contribution of the fermions the difference $a_1 - b_1$ is clearly divergent as Λ is taken to infinity (or arbitrarily large). Also the field renormalization factor Z_H which explicitly contains the cut off dependent term a_1 diverges with rising cut off. However, the main point is, that the divergent parts do not appear anymore with powers of the squared momenta such that all divergences cancelled within the difference of the one particle irreducible terms. From the point of Euclidean lattice field theory one would equivalently phrase that, it is always possible to tune the bare Higgs boson mass m_H and the field renormalization factor Z_H with arbitrary but finite values of the cut off such that the physical quantities are held fixed. The points in the bare parameter space for varying cut off values defines the line of constant physics. Keeping in mind that here the underlying theory renders a trivial theory as the cut off is large enough, one has to restrict the line of constant physics to some finite interval of cut off values.

After this lengthy treatment on the the specific details about the renormalization of the fermionic contribution, the renormalized Higgs boson propagator is given by

$$Z \left(G_H^R \left(p^2, M_H^2, M_G^2, \lambda_R, y_R; \Lambda \right) \right)^{-1} =$$
$$p^2 + M_H^2 + 18 \left(4\lambda_R v \right)^2 \left(I_1(p^2, M_H^2, \Lambda) - I_1(-M_H^2, M_H^2, \Lambda) \right)$$
$$+ 6 \left(4\lambda_R v \right)^2 \left(I_1(p^2, M_G^2, \Lambda) - I_1(-M_H^2, M_G^2, \Lambda) \right)$$
$$- \Delta \Sigma_H^f \left(p^2, M_f = y_r v_R, y_R; \Lambda \right).$$

Figure 3.2 shows the Higgs boson propagator for some arbitrary values of the couplings and the masses. The first image shows a stable Higgs boson. The plot was made by taking an artificial value for the Goldstone mass larger than zero. The imaginary part of the propagator vanishes as long as the squared momentum is smaller than two-Higgs boson energy treshold ($|p^2| < 4M_H^2$). The curve starting at $-p^2 > 4M_H^2$ denotes the imaginary part of the Higgs boson propagator. Generally Goldstone bosons are massless and thus the Higgs boson can decay into any even number of Goldstone bosons. The second image shows the case where the Goldstone mass is taken to zero. The imaginary part arises much earlier than $-p^2 > 4M_H^2$ and the continuous multi-particle states are induced below the pole of the Higgs boson mass.

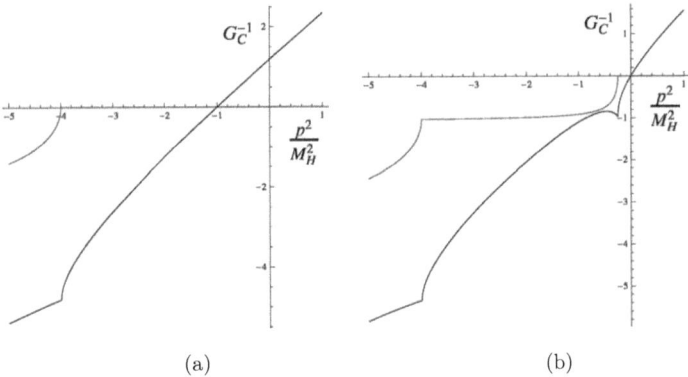

Figure 3.2: The following figure shows the Higgs boson propagator for some arbitrary values of the couplings. The curve which extends to positve values of $\frac{p^2}{M_H^2}$ and which is almost linear for $\frac{p^2}{M_H^2} > 0$ is the real part of the inverse Higgs boson propagator. The other curve denotes the imaginary part. The first image shows a stable Higgs boson at some artificial value for the Goldstone mass larger than zero. The multi-particle states are induced at the threshold value $-p^2 = 4M_H^2$ and the analytic inverse propagator develops a non trivial imaginary part. The second image shows an unstable Higgs boson. The Goldstone mass is chosen to be $M_G = \frac{1}{4}M_H$. The continuous multi-particle states are now induced below the 'pole' of the Higgs boson mass.

3.1.2 The Goldstone boson propagator

The interactions of the Goldstone boson are again inferred from the Lagrangian in (3.3). Table 3.2 shows an overview of the relevant interaction terms and their Wick contractions. The first column identifies the corresponding Feynman graph which is displayed in figure 3.1. The second column shows the coupling in the Lagrangian in (3.3) from which the specific interaction arises while the third column shows the order in the perturbative expansion of the interaction term. The next column shows one of the possible Wick contractions associated to the interaction term and the last column gives the symmetry factor which denotes the number of topologically equivalent Feynman diagrams. In contrary to the Higgs boson mass corrections, the renormalized mass of the Goldstone boson is known from the general Goldstone theorem which is discussed in many books

Table 3.2: The table below shows the relevant interaction terms in the Lagrangian which contribute to the one particle irreducible diagrams of the Goldstone boson propagator. The fermionic contributions are not considered due to the reasoning at the beginning of this section. The first column is an identifier for the corresponding diagrams shown in figure 3.1. S_I is a part of the interaction Lagrangian. The middle column shows the order in the expansion of the exponential of the interaction term. The next column displays a Wick contraction and its multiplicity is given in the last column.

	S_I	Perturbative expansion	Wick contraction type	fac.
A	$\lambda \left(\mathcal{G}^T \mathcal{G}\right)^2$	$-\lambda \left(\mathcal{G}^T \mathcal{G}\right)^2$	$\mathcal{G}_x^T (\mathcal{G}^T \mathcal{G} \, \mathcal{G}^T \mathcal{G})_{x_1} \mathcal{G}_y$	$4 \cdot 3$
B	$2\lambda \left(\mathcal{G}^T \mathcal{G}\right) H^2$	$-2\lambda \left(\mathcal{G}^T \mathcal{G} \, H^2\right)_{x_1}$	$\mathcal{G}_x^T (\mathcal{G}^T \mathcal{G} \, HH)_{x_1} \mathcal{G}_y$	2
G	$4\lambda v \left(\mathcal{G}^T \mathcal{G}\right) H$	$\frac{(4\lambda v)^2}{2} \left(\mathcal{G}^T \mathcal{G} H\right)_{x_1} \cdot \left(\mathcal{G}^T \mathcal{G} H\right)_{x_2}$	$\mathcal{G}_x^T (\mathcal{G}^T \mathcal{G} H)_{x_1} (\mathcal{G}^T \mathcal{G} H)_{x_2} \mathcal{G}_y$	$4 \cdot 2$

about quantum field theory, e.g. [42]. However, for the sake of completeness the Goldstone propagator is discussed roughly.

Compared to the Higgs boson propagator there is one new type of integral arising from the Wick contractions which contains two mass scales $I_2(p^2, m_\varphi^2, m_G^2, \Lambda)$. Furthermore, there are three massless Goldstone bosons which amounts t an overall factor of three. The self energy of the Goldstone boson propagator reads

$$\frac{1}{3}\Sigma_G\left(p^2, m_\varphi^2, m_G^2, \lambda, y; \Lambda\right) = -12\lambda \, I_1\left(p^2, m_G^2, \Lambda\right) - 4\lambda \, I_1\left(p^2, m_\varphi^2, \Lambda\right)$$
$$+ 4 \cdot (4\lambda v)^2 \, I_2\left(p^2, m_\varphi^2, m_G^2, \Lambda\right) + \Sigma_G^f\left(p^2, m_f, y; \Lambda\right).$$

Σ_G^f denotes the fermionic contribution to the Goldstone boson self energy which will not be explicitly given. The Goldstone boson mass is anyway known from the Goldstone theorem and although in the framework of lattice Monte Carlo simulations, the Goldstone bosons are not exactly massless, they are stable particles such that the Euclidean two point correlation function is a reliable tool to extract their masses in finite volume.

In the following the integral I_2 and the renormalization of the Goldstone boson propagator will be discussed. The appendix B contains a detailed calculation of the integrals.

3.1 Perturbative expansion in the continuum

The renormalization condition for the Goldstone bosons shall be

$$G_G^{-1}\left(p^2 = -\mu^2, \zeta, \Lambda\right) \stackrel{!}{=} 0.$$

ζ stands for the collective set of bare parameters

$$\zeta = \left(m_\varphi^2, m_G^2, \lambda, y\right).$$

As mentioned at the beginning of this chapter, the bare Goldstone boson mass m_G will be kept explicitly though its value is known to be zero, in order to be able to compare the results with the propagator obtained from lattice Monte Carlo simulations. Using the expression for the Goldstone boson self energy one gets

$$-\mu^2 + m_G^2 - \Sigma_G\left(-\mu^2, \zeta; \Lambda\right) = 0.$$

Solving the above equation for m_G^2 and substituting m_G^2 in the inverse propagators causes only an error of order λ^4 which is beyond the one loop approximation performed here. The propagator is then given by

$$G_G^{-1}\left(p^2, \zeta, \Lambda\right) = p^2 + \mu^2 - \left(\Sigma_G\left(p^2, \zeta; \Lambda\right) - \Sigma_G\left(-\mu^2, \zeta; \Lambda\right)\right) + \mathcal{O}\left(\lambda^4\right).$$

The difference of the self energy terms is then given by

$$\frac{1}{3}\Delta\Sigma_G\left(p^2, \mu^2, \zeta; \Lambda\right) = \frac{1}{3}\Sigma_G\left(p^2, \zeta; \Lambda\right) - \frac{1}{3}\Sigma_G\left(-\mu^2, \zeta; \Lambda\right)$$

$$= -12\lambda\left(I_1\left(p^2, m_G^2, \Lambda\right) - I_1\left(-\mu^2, m_G^2, \Lambda\right)\right)$$

$$- 4\lambda\left(I_1\left(p^2, m_\varphi^2, \Lambda\right) - I_1\left(-\mu^2, m_\varphi^2, \Lambda\right)\right)$$

$$+ 4(4\lambda v)^2\left(I_2\left(p^2, m_\varphi^2, m_G^2, \Lambda\right) - I_2\left(-\mu^2, m_\varphi^2, m_G^2, \Lambda\right)\right)$$

$$+ \left(\Sigma_G^f\left(p^2, m_f, y; \Lambda\right) - \Sigma_G^f\left(-\mu^2, m_f, y; \Lambda\right)\right).$$

As will be shown shortly, the divergences in the scalar loop integrals $I_{1/2}$ can be eliminated by a redefined bare mass. The integral expression I_1 has already been presented in the last subsection while dealing with the one loop results for the Higgs boson. The integral I_2 is

$$I_2(p^2, m_\varphi^2, m_G^2, \Lambda) = \frac{1}{32\pi^2}\left\{2 + \log\left(\frac{\Lambda^4}{m_G^2 m_\varphi^2}\right) + \frac{\left(m_G^2 - m_\varphi^2\right)}{p^2}\log\left(\frac{m_G^2}{m_\varphi^2}\right)\right.$$

$$\left. + \log\left(\frac{p^2 + m_G^2 + m_\varphi^2 - p^2\,\kappa(p^2) + i\epsilon\,\mathrm{Sgn}\left(p^2\right)}{p^2 + m_G^2 + m_\varphi^2 + p^2\,\kappa(p^2) - i\epsilon\,\mathrm{Sgn}\left(p^2\right)}\right)\kappa(p^2)\right\}.$$

$$\kappa^2(p^2) := \frac{4\,p^2 m_\varphi^2 + \left(p^2 + m_G^2 - m_\varphi^2\right)^2}{p^4}.$$

The terms involving the cut off Λ are independent of the squared momentum and thus the difference

$$I_2\left(p^2, m_\varphi^2, m_G^2, \Lambda\right) - I_2\left(-\mu^2, m_\varphi^2, m_G^2, \Lambda\right)$$

does not depend on the cut off. Neglecting the fermionic contributions, the renormalized Goldstone boson propagator is given by

$$\left(G_G^R\right)^{-1} = p^2 + \mu^2 - \Delta\Sigma_G\left(p^2, \mu^2, \zeta_R; \Lambda\right)$$

where ζ_R stands for the renormalized set of couplings. The analytic form of the above Goldstone propagator will be used to fit the numerical data on the euclidean part obtained by Monte Carlo simulations.

Figure 3.3 shows the Goldstone boson propagator obtained from Monte Carlo simulation. The two plots correspond to different physical situations. The above analytical form is used as a fit function. The left image 3.3a corresponds to a small value of the bare quartic coupling ($\hat{\lambda} = 0.01$). The Yukawa coupling is chosen such that the top quark is about 175 GeV ($\hat{y} = 0.36274$). The parameter κ is 0.12950 and yields a cut off of approximately 900 GeV. One can observe that the tree level fit function (solid-grey) and the one loop result (solid-black) are equally good. The second plot 3.3b shows the Goldstone boson propagator at infinite bare quartic coupling. The Yukawa coupling is chosen such that it corresponds to a heavy 700 GeV ($\hat{y} = 2.2$) quark and the parameter κ is 0.21300 and yields a cut off of about 3500 GeV. Again the solid grey curve corresponds to a fit to the tree level propagator and the solid black curve denotes the fit to the one loop result. Obviously the tree level result is a too crude approximation in order to extract the finite volume Goldstone boson mass and the field renormalization factor Z_G. The right image also shows that even at infinite bare quartic coupling and rather large Yukawa couplings the analytical form is suitable to serve as a good fit function.

3.2 Lattice perturbation theory

The basic principles of perturbation theory as explained at the beginning of this chapter are also valid on a finite discretized space time lattice. Of course one has to expand the

3.2 Lattice perturbation theory

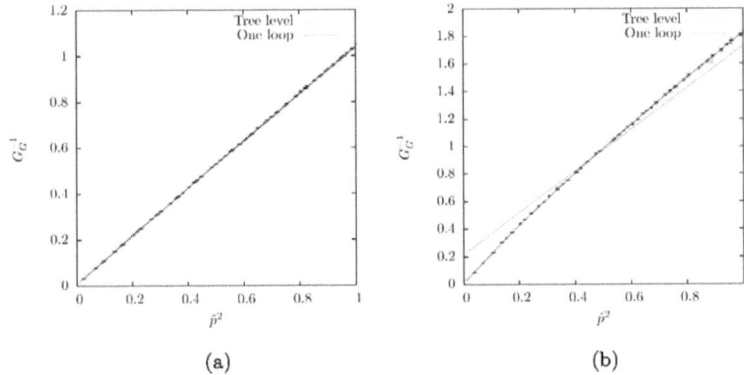

Figure 3.3: The figure shows the Goldstone boson propagator for two different physical situations. The left image corresponds to small bare quartic and Yukawa couplings while the right image corresponds to infinite bare quartic couplings and Yukawa couplings which belong to a quark mass of about 700 GeV. The solid curves are fits of the Monte Carlo data to the analytic form derived above. The grey curve denotes the tree level relation and the black curve belongs to the one loop result. While at small bare couplings, the tree level result is in agreement with the one loop result, the situation changes drastically in the case for large couplings. The right image also shows that even at infinite bare quartic coupling and rather large Yukawa couplings the analytical form is suitable to serve as a good fit function.

lattice action instead of the continuum action given in (2.1). Furthermore, the scalar propagator as well as the fermion propagator is modified.

The main difficulty will be to resemble the coupling structure with the modified chiral projection operators given in the lattice action. Due to the finite set of lattice momenta the integrals are replaced by finite sums which cannot be expressed in a closed formula. The latter also implies a regulator such that no explicit cut off regulator has to be chosen. Finally the sums are evaluated numerically and the perturbative result for the propagator is confronted to those obtained from some selected Monte Carlo simulations.

The Euclidean discretized action is given in (2.7). The observable of interest in this section is the Higgs boson propagator. The scalar interactions lead to the very same diagrams as in the continuum model and hence they will not be discussed in detail. The couplings to the fermions need more care since the modified projection operators do not share the same (anti-) commutator relations as in the continuum.

The fermion action and the fermion matrix were introduced in Chapter 2 equation (2.6)

$$S_F = \sum_{x,y,\alpha,\beta} \begin{pmatrix} \bar{t}_x^\alpha \\ \bar{b}_x^\alpha \end{pmatrix} \mathbb{I} \mathcal{M}_{xy}^{\alpha\beta} \begin{pmatrix} t_y^\beta \\ b_y^\beta \end{pmatrix}$$

$$\mathcal{M}_{xy}^{\alpha\beta} = \left(\mathcal{D}^{(ov)}\right)_{xy}^{\alpha\beta} + \hat{y}\, \Phi_x^\mu \left(P_+^{\alpha\beta}\theta_\mu^\dagger + P_-^{\alpha\beta}\theta_\mu\right)\left(1 - \frac{1}{2}aR\left(\mathcal{D}^{(ov)}\right)_{xy}^{\alpha\beta}\right).$$

As within the framework of perturbation theory in the continuum, it will be assumed that the scalar vacuum expectation value has the form

$$\Phi_x^0 = v + H_x.$$

Isolating the Higgs couplings in the above fermion matrix gives

$$\mathcal{M}_{xy}^{\alpha\beta} = \left(\mathcal{D}^{(ov)}\right)_{xy}^{\alpha\beta} + \hat{y}\left(v + H_x\right)\left(1 - \frac{1}{2}aR\left(\mathcal{D}^{(ov)}\right)_{xy}^{\alpha\beta}\right) + \ldots.$$

As usual in perturbation theory, the free (Gaussian) part of the action will be separated from the rest of the action which defines the interacting part

$$S_F = S_F^0 + S_F^{int}$$

$$S_F^0 := \begin{pmatrix} \bar{t}_x^\alpha \\ \bar{b}_x^\alpha \end{pmatrix} \left\{\left(\mathcal{D}^{(ov)}\right)_{xy}^{\alpha\beta} + \hat{y}v\left(1 - \frac{1}{2}aR\left(\mathcal{D}^{(ov)}\right)_{xy}^{\alpha\beta}\right)\right\} \begin{pmatrix} t_y^\beta \\ b_y^\beta \end{pmatrix}$$

$$S_F^{int} := \hat{y}\, H_x \begin{pmatrix} \bar{t}_x^\alpha \\ \bar{b}_x^\alpha \end{pmatrix} \left(1 - \frac{1}{2}aR\left(\mathcal{D}^{(ov)}\right)_{xy}^{\alpha\beta}\right) \begin{pmatrix} t_y^\beta \\ b_y^\beta \end{pmatrix} + \ldots$$

The interaction with the other scalar fields is left out as only the Higgs boson propagator will be of interest in this chapter. The free fermion propagator is then given by the inverse fermion matrix containing only bilinears in the fermion fields

$$(\Delta_f)_{xy}^{\alpha\beta} := \left\{\frac{1}{\mathcal{D}^{(ov)} + \hat{y}v\left(1 - \frac{1}{2}aR\mathcal{D}^{(ov)}\right)}\right\}_{xy}^{\alpha\beta}.$$

It is useful to introduce the abbreviation $\mathcal{K}$ which is defined by

$$\mathcal{K} := 1 - \frac{1}{2}aR\mathcal{D}^{(ov)}.$$

3.2 Lattice perturbation theory

The contribution of the fermions to the Higgs boson self energy is then given by the following type of Wick contractions

$$G_H^f(x-y;\hat{y}) = -\hat{y}^2\, \mathrm{H}_x \left\{ \left(\overline{\overline{b}}\right)_{x_1}^{\alpha} \mathrm{H}_{x_1} \mathcal{K}_{x_1 x_2}^{\alpha\beta} \left(\overline{\overline{b}}\right)_{x_2}^{\beta} \right\} \left\{ \left(\overline{\overline{b}}\right)_{y_1}^{\sigma} \mathrm{H}_{y_1} \mathcal{K}_{y_1 y_2}^{\sigma\kappa} \left(\overline{\overline{b}}\right)_{y_2}^{\kappa} \right\} \mathrm{H}_y.$$

G_H^f denotes the fermionic contribution the Higgs boson propagator. The above expression is equivalent to

$$G_H^f(x-y;\hat{y}) = -\hat{y}^2 \sum_{x_1,x_2,y_1,y_2} \frac{1}{\Omega}\sum_{q_1} e^{iq_1(x-x_1)} \Delta_{q_1}$$
$$\frac{1}{\Omega}\sum_{k_1} e^{ik_1(x_1-x_2)} \mathcal{K}_{k_1}^{\alpha\beta} \frac{1}{\Omega}\sum_{q_2} e^{iq_2(x_2-y_1)} \Delta_{f,q_2}^{\beta\sigma}$$
$$\frac{1}{\Omega}\sum_{k_2} e^{ik_2(y_1-y_2)} \mathcal{K}_{k_2}^{\sigma\kappa} \frac{1}{\Omega}\sum_{q_3} e^{iq_3(y_2-x_1)} \Delta_{f,q_3}^{\kappa\alpha} \frac{1}{\Omega}\sum_{q_4} e^{iq_4(y_1-y)} \Delta_{q_4}.$$

The sums over the position variables x_1, x_2, y_1, y_2 can be evaluated such that

$$G_H^f(x-y;\hat{y}) = -\hat{y}^2 \frac{1}{\Omega^2} \sum_{q_1,q_4} e^{iq_1 x - iq_4 y} \Delta_{q_1}\Delta_{q_4} \sum_{k_1,q_2} \mathcal{K}_{k_1}^{\alpha\beta} \Delta_{f,q_2}^{\beta\sigma}$$
$$\sum_{k_2,q_3} \mathcal{K}_{k_2}^{\sigma\kappa} \Delta_{f,q_3}^{\kappa\alpha} \delta_{k_1,q_1+q_3} \delta_{k_1,q_2} \delta_{k_2,q_2-q_4} \delta_{q_3,k_2}.$$

Finally the sums over q_1, k_1, q_2, k_2 can be performed by using the Kronecker δ

$$G_H^f(x-y;\hat{y}) = -\hat{y}^2 \frac{1}{\Omega} \sum_{q_1} e^{iq_1(x-y)} \Delta_{q_1} \Delta_{q_1} \sum_{q_3} \mathcal{K}_{q_1+q_3}^{\alpha\beta} \Delta_{f,q_1+q_3}^{\beta\sigma} \mathcal{K}_{q_3}^{\sigma\kappa} \Delta_{f,q_3}^{\kappa\alpha}$$
$$\Rightarrow \tilde{G}_H^f(p;\hat{y}) = -\hat{y}^2 \Delta_p \Delta_p \sum_q \mathcal{K}_{p+q}^{\alpha\beta} \Delta_{f,p+q}^{\beta\sigma} \mathcal{K}_q^{\sigma\kappa} \Delta_{f,q}^{\kappa\alpha}.$$

The final expression describes a fermion loop in lattice momenta. The amputated two point Green function is then obtained by cancelling the scalar Higgs boson propagators. The above expression is the analogue to the corresponding fermion loop in the continuum and there are no new type of interactions which arise from the discretized action.

The expression for the self energy contributions arising from the fermions is then given by

$$\Sigma_H^f(\hat{p},\hat{y},m_f) = -\hat{y}^2 \sum_q \mathcal{K}_{p+q}^{\alpha\beta} \Delta_{f,p+q}^{\beta\sigma} \mathcal{K}_q^{\sigma\kappa} \Delta_{f,q}^{\kappa\alpha}$$
$$= -\hat{y}^2 \sum_q \mathrm{Tr}\left\{\mathcal{K}_{p+q}\Delta_{f,p+q}\,\mathcal{K}_q\Delta_{f,q}\right\}$$
$$m_f = \hat{y}v.$$

The final expression is

$$\Sigma_H^f(\hat{p},\hat{y},m_f) =$$
$$-\hat{y}^2 \sum_q \text{Tr}\left\{ \left(1 - \frac{1}{2}a R\mathcal{D}^{(ov)}\right)_{p+q} \left(\frac{1}{\mathcal{D}^{(ov)} + m_f\left(1 - \frac{1}{2}a R\mathcal{D}^{(ov)}\right)}\right)_{p+q}\right.$$
$$\left. \left(1 - \frac{1}{2}a R\mathcal{D}^{(ov)}\right)_q \left(\frac{1}{\mathcal{D}^{(ov)} + m_f\left(1 - \frac{1}{2}a R\mathcal{D}^{(ov)}\right)}\right)_q \right\}. \quad (3.8)$$

It is the aim to compare the results from the one loop perturbation theory to results obtained by Monte Carlo simulations. In the case where the bare quartic coupling vanishes, the above fermion loop is the dominant contribution to the Higgs boson mass. Figure 3.4 shows a comparison between the perturbative results and the Monte Carlo result on the Higgs boson propagator. The bare quartic coupling is chosen to be zero and the cut off is —GEV520 for the first image and 440 GeV for the other two. The three images correspond to three different Yukawa couplings while the cut off was aimed to be constant. The left image corresponds to $y_0 = 0.35382$, the middle plot corresponds to $y_0 = 0.71139$ generates a physical top quark mass of 175 GeV. Finally the last image was generated at a Yukawa coupling of $y_0 = 1.45192$ and corresponds to a physical top quark mass of about 350 GeV. About 100 000 configurations were produced for each set of parameters in order to reduce the statistical errors. The figure shows an excellent agreement between one loop perturbation theory and the numerical data obtained from Monte Carlo simulations.

The scalar vacuum expectation value can either be calculated by taking the general form of the action (2.7) and assume an arbitrary value for the scalar vev. In general, the vacuum expectation value of the Goldstone bosons and Higgs bosons do not vanish and the tadpole diagrams are the leading contributions to their vacuum expectation value. Up to the considered order in perturbation theory it is then fine to tune the vev such that the tadpole diagrams vanish. An alternative approach was proposed in [17] which does not depend on the phase of the theory. It is based on the effective potential where the classical minimum of the potential yields the ground state and hence the vacuum expectation value. The effective potential is based on an expansion in loops instead of weak couplings and it turns out that the effective potential is much more reliable in predicting the scalar vacuum expectation value than the analysis of tadpole diagrams.

3.2 Lattice perturbation theory

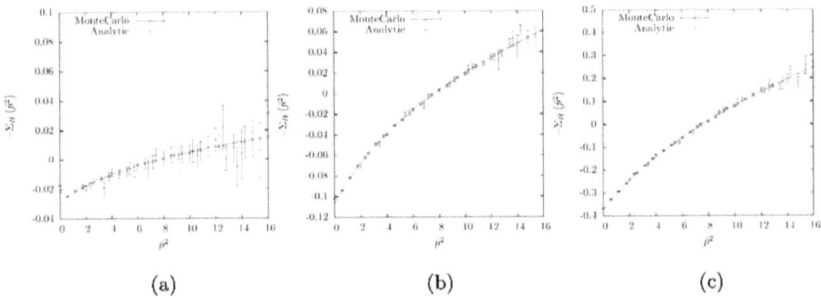

Figure 3.4: The figure shows the results obtained from one loop perturbation theory at vanishing bare quartic coupling contrasted with results obtained from Monte Carlo simulation. Three values of the Yukawa coupling are chosen $\hat{y} \in \{0.17644, 0.35288, 0.70576\}$. κ tuned in order to keep the cut off in the three cases fixed ($\kappa \in \{0.12434, 0.12303, 0.11814\}$). The images are ordered from left to right with rising Yukawa coupling. The calculations where performed on a 8^4 lattices.

The effective potential in the Higgs-Yukawa model was investigated in [25] and will not be presented here.

Finally, the dependence of the parameter ρ in the definition of the Neuberger operator (2.5) on the Higgs boson propagator is shown in figure 3.5. The plots are obtained with the help of the perturbative one loop result. Again the effect of the fermions should be dominant in the case where the quartic bare coupling vanishes. The numerical data corresponds to $\rho = 1$ and is in very good agreement with the analytical data (solid black curve) at $\rho = 1$. Figure 3.5 shows three different values of ρ which were computed within perturbation theory. The final plot shows the summary of all chosen values of $\rho \in \{0.5, 0.75, 1.0\}$. The data in figure 3.5 were performed on a 8^4 lattice which is two small in order to obtain physical results. Nevertheless, it turns out that there is a strong dependence of the Higgs boson self energy on the parameter ρ. From the latter one infers that for all three values of the parameter ρ, one has to adjust the bare parameters of the theory such that Higgs boson propagator is evaluated at the same value of the cut off.

59

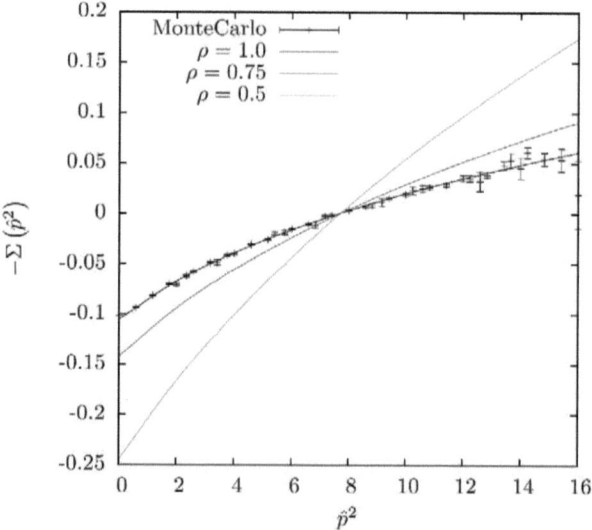

Figure 3.5: The figure shows the dependence of the Higgs boson propagator on the parameter ρ of the Neuberger operator. The Yukawa coupling is $\hat{y} = 0.35288$ and corresponds to the middle plot in figure 3.4. The calculations where performed on a 8^4 lattice.

4 Resonance parameters of the Higgs boson

This standard model Higgs boson as well as the Higgs boson in the Higgs-Yukawa model considered here, is not a stable particle. If the Higgs boson mass is larger than twice the weak gauge boson masses ($2M_W^\pm$ or $2M_Z$), it will decay and only the weak gauge bosons are in the spectrum of the asymptotic theory. The same applies for the pure Higgs Yukawa model, where the Higgs boson decays into any even number of Goldstone bosons. In fact the Goldstone bosons are massless particles and thus the Higgs boson is never in the spectrum of the asymptotic theory. In the following discussion however, the Goldstone bosons will be treated as generic scalar particles with explicit mass provided by a coupling to an external source.

Figure 4.1 shows the current experimental status on Higgs boson searches at LHC and Tevatron [44]. The image shows the $\Delta\chi^2 = \chi^2 - \chi^2_{min}$ against the Higgs boson mass. The shaded area up to 114 GeV denotes the excluded mass ranges from direct Higgs boson searches at LEP-II and Tevatron (158 GeV to 175 GeV). The solid line represents a fit of the standard model to high precision electroweak measurements and shows that the preferred value of 89^{35}_{26} GeV is already excluded.

As discussed in Chapter 2, a signature of the decaying particle is left in the propagator after it has been analytically continued into the complex plane. The physical Higgs boson propagator, which is always real valued, exhibits a branch cut at the threshold energy for the production of two weak gauge bosons or respectively two Goldstone bosons. In the case, where the Higgs boson is stable, the Higgs boson propagator will have a pole before the branch cut which identifies the physical Higgs boson mass. In the unstable case, the Higgs boson propagator still has the branch cut, but no pole before the branch cut.

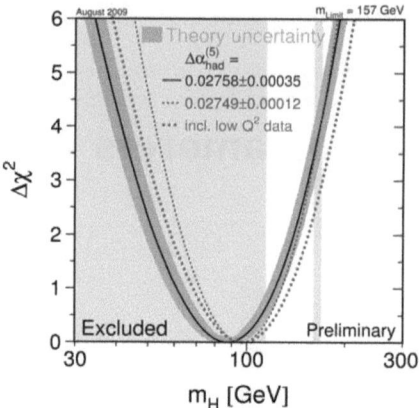

Figure 4.1: The figure shows the experimental exclusion limit on the Higgs boson mass [44]. The vertical axis shows $\Delta\chi^2 = \chi^2 - \chi^2_{min}$ and the horizontal axis denotes the Higgs boson mass. The shaded range update about 114 GeV shows the excluded mass range from direct Higgs boson searches at LEP-II and Tevatron. The solid curve is a fit of the standard model to electroweak precision measurements while the dark grey band alogn the solid curve corresponds to the theoretic uncertainty which is associated to higher order calculations in perturbation theory.

In order to characterize the unstable Higgs boson, it is necessary to consider the analytic continuation of the physical propagator to the complex plane. The analytic continuation was discussed in Chapter 2 in detail. The physical propagator will be denoted by $G : \mathbb{R} \to \mathbb{R}$ and its analytic continuation shall be $G_{\mathbb{C}} : \mathbb{C} \to \mathbb{C}$. The physical propagator is then obtained by taking the limit of $G_{\mathbb{C}}(z)$ where z approaches the real axis from below. As discussed in Chapter 2, the analytic continuation has no discontinuity along the branch cut and the physical Higgs boson mass is given by locating the complex pole in the second Riemann sheet.

The existence of an unstable particle though, is revealed through a "resonant" behaviour of the total cross section in the two body scattering of the corresponding asymptotic final states. In the following, it is the aim to clarify the role of resonances in cross sections and to roughly sketch the connection between the two point Green function and the so-called scattering phase.

This chapter starts with the current mass bounds obtained from numerical simulations. For the upper Higgs boson mass bound, where the Higgs boson decays into Goldstone bosons, an analytic structure of the Higgs boson propagator had to be assumed. Afterwards the connection of the total cross section and the resonance mass as well as the resonance width is discussed in continuum field theory. An analogous relation in finite volume was worked out by Lüscher [46] and the method is briefly explained. Finally the numerical results for the scattering phase are presented. The obtained resonance parameters are compared to those obtained by means of the Higgs boson propagator.

4.1 Mass bounds of the Higgs boson

The current upper and lower Higgs boson mass bounds, which were obtained from lattice simulations in the Higgs-Yukawa model described in chapter 2, were established within an extensive investigation of Euclidean two point correlators (see [31, 30]). In the case of the lower bound, the Higgs boson turns out to be a stable particle and its mass is given by the pure ground state energy. These can be accessed with the help the time correlators. The analysis of the upper Higgs boson mass involves an unstable Higgs boson, which decays into Goldstone bosons. In order to access the unphysical propagator G_C a functional form of the propagator which is motivated by perturbation theory was assumed. The real part of the pole in the second Riemann sheet was located after a fit of the propagator to the numerical data. The same procedure can be performed for the lower Higgs boson mass, where the Higgs boson is a stable particle and the results are consistent with those obtained from the Euclidean two point correlation functions.

Nevertheless, a genuinely non perturbative analysis of the resonance parameters is desirable as one cannot know, whether the functional form of the propagator may receive severe contributions at larger values of the renormalized coupling, which is not any more reflected in the one loop approximation. Furthermore, it not clear whether it is sufficient to identify the Higgs boson mass by locating the zero of the real part of the propagator instead the complex pole.

4.1.1 Observables and Higgs boson mass bounds

This section introduces the basic observables and gives some details on the determination of mass parameters in the framework of Euclidean lattice field theory. Finally the current numerical bounds for the upper and lower Higgs boson mass are discussed.

The definitions of the Higgs and the Goldstone boson fields were given in (3.2) and are repeated here

$$\varphi = \begin{pmatrix} \mathcal{G}_1 + i\mathcal{G}_2 \\ v + H + i\mathcal{G}_3 \end{pmatrix}.$$

v is a constant and corresponds to the scalar vacuum expectation value. Within the Higgs-Yukawa model it is connected with the magnetization (see (2.9) and (2.10))

$$\text{mag} := |\overline{\Phi}| = \left(\sum_\alpha \overline{\Phi}_\alpha^2\right)^{\frac{1}{2}}, \qquad \overline{\Phi}_\alpha = \frac{1}{V}\sum_{x\in\mathbb{Z}_L^4} \Phi_\alpha(x)$$

$$v := \sqrt{2\kappa}\,\text{mag}$$

$$\Rightarrow v_R := \sqrt{2\kappa}\,\frac{\text{mag}}{\sqrt{Z_G}}.$$

Z_G is the field renormalization factor of the Goldstone fields. The physical vev (v_R) is known to be 246 GeV. The simluation strategy and the tuning of the bare parameters is discussed in section 2.3. The basic definitions of the Higgs and Goldstone boson masses are given in Chapter 2 equation (2.15).

Goldstone bosons are almost massless on the lattice with a mass of $m_G \propto \frac{1}{L^2}$. Therefore, the Higgs boson mass receives particular finite size effects of the order $\mathcal{O}(\frac{1}{L^2})$ which does not follow the usual exponentially suppressed finite size effect given by $\mathcal{O}\left(e^{-mL}\right)$. An infinite volume extrapolation is therefore inevitable. Furthermore, the masses in lattice units must be significantly smaller than the cut off in order to avoid too large discretization effects. At the same time the finite volume must be large enough such that the compton wave length of the Higgs boson and the fermions are smaller than the lattice extent. The above constraints are summarized by

$$\hat{m} \cdot L_{s,t} > 2, \qquad \hat{m} < \frac{1}{2}\Lambda.$$

L_s denotes the spatial extent and L_t corresponds to the temporal extent of the lattice. The mass $\hat{m}$ refers to the mass in lattice units.

The upper and the lower Higgs boson mass bounds have been investigated earlier. The results were obtained mainly within the $\lambda\phi^4$ model. An analysis of the Higgs resonance in the $\lambda\phi^4$ at infinite bare quartic coupling is published in [35]. The model considered in this work involves the coupling of the Higgs boson to the fermion fields and incorporates all dynamics of the fermionic degrees of freedom. The model is therefore more realistic with respect to the standard model of particle physics. It is the aim to compare the conceptual aspects of extracting the mass of an unstable particle and thus only those mass bounds will be considered which were obtained within the pure Higgs-Yukawa model using overlap fermions.

The lower Higgs boson mass is determined at vanishing bare quartic coupling and thus the resonance width is expected to be very small. It is then sufficient to determine the zero of the real part of the Higgs boson propagator in order to compute the Higgs boson mass. Furthermore, the time dependence of the Higgs boson correlator is dominated by the physical Higgs boson mass. However, the finite size effects have to be investigated and finally an extrapolation to infinite volume has to be performed. The results for the lower Higgs boson mass together with the finite size analysis is presented in figure 4.2 and was published in [30]. The figure summarizes the volume dependence of the Higgs boson mass and shows the obtained infinite volume extrapolation of the Higgs boson masses. The image in figure 4.2b compares the obtained infinite volume results in the case of a degenerate quark doublet with results obtained from the effective potential at different values of the number of fermion doublets and also for non degenerate quark doublets. The numerical data agrees very well with the perturbative prediction with a single degenerate fermion doublet ($N_f = 1$, $\frac{y_b}{y_t} = 1$). The solid line represents the physical situation with three generations of fermions and a mass splitting within the fermion doublet ($\frac{y_b}{y_t} = 0.024$).

The upper Higgs boson mass suffers from the conceptual difficulties which were already mentioned at the beginning of this chapter and which are discussed in more detail in Chapter 2.3. For the sake of completeness the basic procedure is roughly summarized in the following.

An analytic expression for the Higgs boson propagator is suggested by renormalized

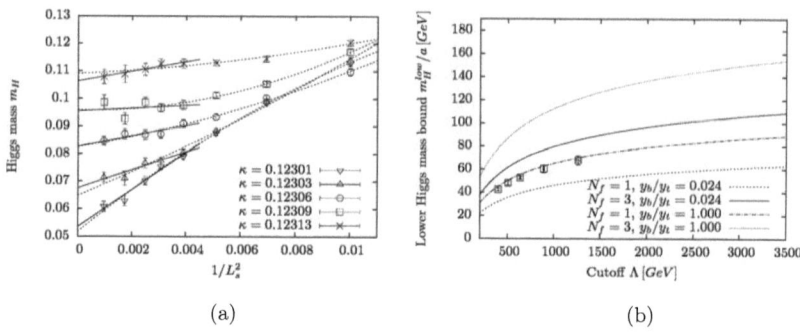

Figure 4.2: The figure shows the published bounds for the lower Higgs boson mass. Image (a) shows the volume dependence of the Higgs boson mass together with a linear and a quadratic extrapolation to infinite volume. Image (b) shows the final result of the lower Higgs boson mass. The dashed curves are perturbative results obtained from the analysis of the effective potential. The numerical data agrees very well with the perturbative prediction with a single degenerate fermion doublet ($N_f = 1$, $\frac{y_b}{y_t} = 1$). The solid line represents the physical situation with three generations of fermions and a mass splitting within the fermion doublet ($\frac{y_b}{y_t} = 0.024$).

perturbation theory. The inverse renormalized Higgs boson propagator is then given by

$$\left(G_H^R\right)^{-1}\left(p^2, M_H^2, \ldots\right) = p^2 + M_H^2 - \Sigma_H\left(p^2, M_H^2, \ldots\right)$$

(the dots indicate the dependence on the parameters and the cut off of the model). M_H denotes the 'physical' Higgs boson mass. The physical propagator is real valued and a fit to the lattice propagator data can be performed where the physical Higgs boson mass and the renormalized quartic coupling are taken as free fit parameters. It was sufficient to compute the inverse Higgs boson propagator up to on loop in the scalar fields. The contribution of the fermion loop was neglected. However, the field renormalization factor Z_H was taken into account although Z_H does not receive any contributions from the scalar one loop approximation. The leading order contribution to the scalar field renormalization factor arises from the fermion loop. The upper Higgs boson mass bound has been published in [31] and is shown in figure 4.3.

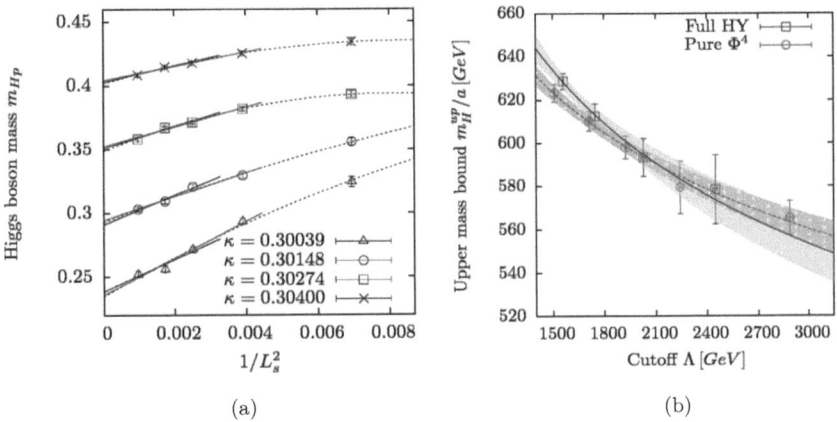

Figure 4.3: The left image (a) shows the finite size effects of the Higgs boson mass and an extrapolation to infinite volume, The results are summarized in the final plot (b). The shaded region indicates the maximal and the minimal value of the upper Higgs boson mass bound which is attainable within the standard deviations of the fit parameters. Furthermore, the right image (b) displays the upper Higgs boson mass in dependence of the cut off Λ in the pure $\lambda\phi^4$ theory and in the Higgs-Yukawa model involving dynamical overlap fermions. The results are taken from [31].

4.2 Resonance mass and width of the Higgs boson

Though the established numerical results presented in [31, 30] agree with analytically expected curve, it is based on a one loop result of the Higgs boson propagator. Furthermore, the pole of the Higgs boson was identified on the real p^2 axis, though a resonance is in general displaced from the real axis. In the case where the width Γ is small with respect to the physical resonance mass, the above restriction to the real axis is a good approximation. A priori, the resonance width and the mass is unknown and a genuinely non perturbative analysis of the resonance parameters is desirable in order to confirm or disprove the above approach.

The method to compute the resonance parameters in finite volume was proposed in [46] and it has been successfully employed in QCD and in the pure $\lambda\phi^4$ theory [35, 19].

From the unitarity of the S-matrix, many aspects of resonances and analyticity properties of the Green functions can be determined independent of the underlying interactions. The basic quantity of interest will be the forward scattering amplitude which is parametrized by the total centre of mass energy. The unitarity condition of the S-matrix then yields a relation between the forward scattering amplitude and the centre of mass energy which can be solved by introducing a phase, the so called scattering phase.

In the following, the scattering phase in continuum field theory will be discussed. Afterwards the main aspects of the analogous relation on a finite, discretized lattice will be elaborated. The latter is a short summary of the detailed calculations given in [46].

4.2.1 The scattering phase in the continuum and in a finite box

The unitarity of the S-matrix is one of the fundamental prerequisites in quantum field theory. It reflects the conservation of probabilities for transitions between asymptotic states.

The short discussion below on the scattering phases in continuum quantum field theory follows the detailed presentation in [56, 42].

The S-matrix can be decomposed into the free part and the part which is responsible for interactions

$$S = \mathbb{I} + iT.$$

The unitarity condition then yields

$$S^\dagger S = \mathbb{I}$$
$$\Rightarrow T^\dagger T = -i\left(T - T^\dagger\right).$$

The matrix expression can be written in its matrix elements by projecting onto asymptotic states

$$\langle f | T | i \rangle = (2\pi)^4 \delta(P_f - P_i) T_{fi}. \tag{4.1}$$

where P_f, P_i is the total momentum in the final or initial state respectively. Written in matrix elements, the unitarity for the interacting part of the S-matrix reads

$$T_{fi} - T_{if}^* = i \sum_n (2\pi)^4 \delta^4 (P_n - P_i) T_{nf}^* T_{ni}. \tag{4.2}$$

4.2 Resonance mass and width of the Higgs boson

In the last step a complete set of states has been inserted in (4.1). The sum contains a finite number of one particle states and an integral over a continuous spectrum of states. In order to ease the notation the sum shall represent the finite one particle states as well as the integral over the continuous spectrum of states. In the special case of forward scattering, the final state is identical to the initial state and the right hand side of equation (4.2) is related to the total cross section. Equation (4.2) then reduces to

$$\Im\{T_{ii}\} = \lambda^{\frac{1}{2}}\left(s, m_a^2, m_b^2\right)\sigma_{tot}(i).$$

λ is kinematical factor which is completely determined by the initial state variables

$$\lambda \equiv \lambda(s, m_a^2, m_b^2) := \left(s^2 + m_a^4 + m_b^4\right) - 2s\,m_a^2 - 2s\,m_b^2 - 2\,m_a^2 m_b^2.$$

s denotes the total centre of mass energy of two incoming scalar particles $s = (p_a + p_b)^2$. Furthermore, the matrix elements of T can be expanded in partial waves

$$T_{fi} = 16\pi \sum_J (2J+1) P^J(\cos\theta)\, T_{fi}^J(s).$$

P^J denotes the Jth Legendre polynom. The above relation holds for spinless particles. If the total energy $\sqrt{s}$ is below the inelastic threshold, the sum over the complete set of states (4.2) reduces to the two particle states. The projection to the Jth angular momentum part yields

$$T^J(s) - T^{J*}(s) = 2i\frac{\sqrt{\lambda}}{s} T^{J\dagger}(s) T^J(s).$$

The above equation is solved by

$$\frac{2\sqrt{\lambda}}{s} T^J(s) = -i\left(e^{2i\delta_J(s)} - 1\right) = 2\,e^{i\delta_J(s)}\,\sin\delta_J(s). \tag{4.3}$$

δ_J is called the scattering phase. Using the optical theorem, which is a consequence of the unitarity of the S-matrix and does not depend on the details of the interaction, the left hand side of equation (4.3) can be associated with the total cross section of the two particle initial state $|i\rangle$.

$$\sigma_{tot} = \frac{8\pi}{q\sqrt{s}} \sum_{J=0}^{\infty} (2J+1)\,\Im\{T^J\} \tag{4.4}$$

q is the centre of mass momentum. It must be stressed that the above equation is only valid for energies below the inelastic threshold. In the vicinity of a resonance, the total cross section will exhibit a peak which resembles the Breit-Wigner curve. The left hand side can thus be parametrized with the resonance mass and the resonance width. It is only valid if the energy E is close to the resonance mass (see [53] or (2.14))

$$\sigma_{tot} = \left| \frac{1}{p^2 - m^2 + im\Gamma} \right|^2 \quad (4.5)$$

where m is the resonance mass and Γ is the resonance width. The final relation (4.3), (4.4), and (4.5) connects the energy of the two particle system with the scattering phase δ. Accordingly it can be used to determine the resonance mass, if the scattering phase is known. An analogous relation in finite volume will be presented below.

The main result in Lüscher's work [46] is the connection between the two particle energies in a finite box and the scattering phase in infinite volume. Once the scattering phases are determined through the analysis of the two particle energies in lattice simulations, it will allow to compute the resonance mass and width of unstable particles. A brief overview of the method is given below. The main idea is reduced to the special case of interest here. Within the Higgs-Yukawa model, the Higgs boson dominantly decays into Goldstone bosons. The two particle state of interest is thus a singlet which corresponds to the A^+ symmetry sector. The representations of the cubic symmetry $O(3, \mathbb{Z})$ on the lattice can be derived from the irreducible representation of $O(3)$. Equation (4.14) shows a decomposition in in harmonic polynomials. The A^+ representation corresponds to the trivial representation and belongs to vanishing angular momentum.

In order to express the above relation between the scattering phases and the energy level in a finite box, it is sufficient to investigate the scattering phase in non relativistic quantum mechanics. The non relativistic result can be transferred to the case of quantum field theory [49, 45].

Consider therefore a non-relativistic system in infinite volume. The Hamiltonian shall be given by a potential which has a finite range

$$\mathcal{H} = -\frac{1}{2\mu} \Delta + V(r), \quad r = |\vec{r}|, \quad V(r) \equiv 0 \text{ if } r > R > 0. \quad (4.6)$$

The potential shall be invariant under rotations. The wave functions can then be expanded in spherical harmonics and the coefficients then solve the radial Schrödinger equation. Far away from the potential $r > R$ the radial part of the wave function is then given by a superposition of spherical Bessel and Neumann functions

$$\psi(\vec{r}) = \sum_{l=0}^{\infty} \sum_{m=-l}^{l} Y_{lm}(\theta, \varphi) \, \psi_{lm}(r),$$

$$\psi_{lm}(r) \propto \alpha_l(k) \, j_l(kr) + \beta_l(k) \, n_l(kr), \quad E = \frac{k^2}{2\mu}.$$

Y_{lm} denotes the spherical harmonics which are solutions of the angular part of the Laplace operator in (4.6). j_l and n_l are the spherical Bessel and Neumann functions. For real and positive values of k, the scattering phase is defined by

$$e^{2i\delta_l(k)} = \frac{\alpha_l(k) + i\beta_l(k)}{\alpha_l(k) - i\beta(k)}. \tag{4.7}$$

The above steps can be found in many text books on quantum mechanics (see e.g. [52]).

The main task is now to derive an expression in finite but continuous space time volume. After defining the underlying Hamiltonian, the main steps shall be summarized. In a box of size L^3, the wave functions are assumed to fulfil periodic boundary conditions

$$\psi(\vec{r} + \vec{n}L) = \psi(\vec{r}), \quad \forall \vec{n} \in \mathbb{Z}^3$$

$$\mathcal{H} = -\frac{1}{2\mu}\Delta + V_L(\vec{r}), \quad V_L(\vec{r}) := \sum_{\vec{n} \in \mathbb{Z}^3} V(|\vec{r} + \vec{n}L|).$$

It can be shown that, far away from the range of the potential, the radial part of the wave function can still be written as

$$\psi_{lm} \propto \{\alpha_l(k) \, j_l(kr) + \beta_l \, n_l(kr)\}. \tag{4.8}$$

The asymptotic form of the wave function is also preserved if one introduces a finite angular momentum cut off Λ. Though the spectrum of the cut off Hamiltonian $\mathcal{H}_\Lambda$ is different from the original Hamiltonian (4.6), it can be shown that they approach each other in the limit $\Lambda \to \infty$. The cut off Hamiltonian is given by

$$\mathcal{H}_\Lambda := -\frac{1}{2\mu}\Delta + Q_\Lambda V(r)$$

$$Q_\Lambda := \sum_{l=0}^{\Lambda} \sum_{m=-l}^{l} Y_{lm}(\theta, \varphi) \, \psi_{lm}(r).$$

Out of the range of the potential the eigenvalue equation then reduces to the Helmholtz equations

$$\mathcal{H}_\Lambda \psi(\vec{r}) = E\psi(\vec{r})$$

$$\Rightarrow \left(\Delta + k^2\right) \psi(\vec{r}) = 0.$$

In order to keep the main arguments as transparent as possible, the main steps will be enumerated below

1. Construct all singular periodic solutions of the Helmholtz equation.

2. Decompose the Green functions in spherical harmonics.

3. The energy spectrum is given by the asymptotic boundary conditions (4.8).

4. Restrict to spinless scalar particles.

5. The relation between the scattering phase and the two particle energy follows from the zero of a determinant.

The crucial points of the above steps shall be briefly discussed. A complete and detailed demonstration of the steps is given in [46].

1. **Singular periodic solutions of the Helmholtz equation**

   If a solution in the exterior region Ω is given

   $$\Omega = \left\{ \vec{r} \in \mathbb{R}^3 : |\vec{r} + \vec{n}L| > R, \quad \forall \vec{n} \in \mathbb{Z}^3 \right\}$$

   such that its spherical components ψ_{lm} satisfies (4.8), it can be guaranteed that there is a unique eigenfunction of $\mathcal{H}_\Lambda$ which coincides with the above solution in Ω. Within a finite box, the possible momenta are discrete and given by

   $$\mathcal{P}_L := \left\{ \vec{p} \in \mathbb{R}^3 | \vec{p} = \pm \frac{2\pi}{L} \vec{n}, \quad \vec{n} \in \mathbb{Z}^3 \right\}. \quad (4.9)$$

   In the following, it will be assumed that the parameter k^2 in the Helmholtz equation is not in the above set of lattice momenta. The case where k is in the above set must be treated seperately but does not change the final results (see section 4 in [46]).

   The Green function of the Helmholtz equation are solutions of

   $$\left(\Delta + k^2 \right) G(\vec{r}, k^2) = - \sum_{\vec{n} \in \mathbb{Z}^3} \delta \left(\vec{r} + \vec{n}L \right).$$

Using the Fourier transform $G(\vec{r}, k^2) = \frac{1}{L^3} \sum_{p \in \mathcal{P}_L} e^{i\vec{p}\vec{r}}$ one easily finds

$$G(\vec{r}, k^2) = \frac{1}{L^3} \sum_{p \in \mathcal{P}_L} \frac{e^{i\vec{p}\vec{r}}}{p^2 - k^2}.$$

On the other hand the Neumann function obeys

$$\left(\Delta + k^2\right) n_0(kr) = -\frac{-4\pi}{k} \delta(\vec{r}).$$

With the help of the Neumann function, the Green function can be decomposed into a singular and a regular part

$$G(\vec{r}, k^2) = \frac{k}{4\pi} n_0(kr) + \hat{G}(\vec{r}, k^2).$$

It is clear that derivatives with respect to $\vec{r}$ acting on the above Green function yields further singular solutions of the Helmholtz equation. Not all of them are linear independent as one can always use the Helmholtz equation $\Delta \psi(\vec{r}) = -k^2 \psi(\vec{r})$ to map the second order derivative of the Green function to the Green function itself. All linear independent solutions are obtained with the help of the harmonic polynomials

$$\mathcal{Y}_{lm}(\vec{r}) := r^l Y_{lm}(\theta, \varphi)$$

$$\Rightarrow G_{lm}(\vec{r}, k^2) := \mathcal{Y}_{lm}(\nabla) G(\vec{r}, k^2).$$

The solutions G_{lm} are also decomposed in its singular and its regular part

$$G_{lm}(\vec{r}, k^2) = \frac{-1^l}{4\pi} Y_{lm}(\theta, \varphi) k^{l+1} n_l(kr) + \hat{G}_{lm}(\vec{r}, k^2).$$

2. **Expansion in spherical harmonics**

A generic singular solution is now given as a linear combination of the above Green functions G, G_{lm}. The expansion of the regular part in yields

$$G(\vec{r}, k^2) = \frac{k}{4\pi} n_0(kr) + \sum_{l=0}^{\infty} \sum_{m=-l}^{l} g_{lm} Y_{lm}(\theta, \varphi) j_l(kr) \tag{4.10}$$

$$G_{lm}(\vec{r}, k^2) = \frac{-1^l k^{l+1}}{4\pi} \left\{ Y_{lm}(\theta, \varphi) n_l(kr) + \sum_{j=0}^{\infty} \sum_{s=-j}^{j} \mathcal{M}_{lm,js} Y_{js}(\theta, \varphi) j_j(kr) \right\}. \tag{4.11}$$

In the case (4.10) the coefficients g_{lm} are given by a generalized Zeta function

$$g_{lm} = \frac{i^l}{\pi q^l L} \mathcal{Z}_{lm}(1; q^2), \qquad q = \frac{kL}{2\pi}$$

$$\mathcal{Z}_{lm}(s; q^2) := \sum_{\vec{n} \in \mathbb{Z}^3} \mathcal{Y}_{lm}(\vec{n}) \frac{1}{(n^2 - q^2)^s}.$$

The special case where $l = 0$ and $m = 0$ will be relevant for the Higgs boson resonance. The above equations can be simplified considerably when restricted to this case

$$g_{00} = \frac{1}{\pi L} \mathcal{Z}_{00}\left(1; q^2\right)$$
$$= \frac{1}{\pi L} \sum_{\vec{n} \in \mathbb{Z}^3} \frac{1}{\sqrt{4\pi}} \frac{1}{n^2 - q^2}.$$

3. **Energy eigenstates**

From equation (4.8) one infers that the general solution of the spherical components is given by

$$\psi_{lm}(r) = b_{lm} \left\{ \alpha_l(k) \, j_l(kr) + \beta_l \, n_l(kr) \right\}. \tag{4.12}$$

At the same time, the general singular periodic solution of the Helmholtz equation with degree Λ can be represented as a linear combination of the Green functions G_{lm}

$$\psi(r) := \sum_{l=0}^{\Lambda} \sum_{m=-l}^{l} v_{lm} G_{lm}(\vec{r}, k^2).$$

Inserting the expansion (4.11) into (4.12) yields

$$b_{lm}\alpha_l(k) = \sum_{j=0}^{\Lambda} \sum_{s=-j}^{j} v_{js} \frac{-1^j}{4\pi} k^{j+1} \mathcal{M}_{js,lm}$$

$$b_{lm}\beta_l(k) = v_{lm} \frac{-1^l}{4\pi} k^{l+1}.$$

The second equation determines v_{lm}. The first equation is then

$$b_{lm}\alpha_l(k) = \sum_{j=0}^{\Lambda} \sum_{s=-j}^{j} b_{js}\beta_j(k) \mathcal{M}_{js,lm}. \tag{4.13}$$

In order to express the above relation as a matrix equation one defines operators in the vector space of elements $b_{lm}, l \in \{0, \ldots \Lambda\}, m \in \{-l, l\}$

$$A_{lm,js} := \alpha_l(k) \, \delta_{lj} \delta_{ms}, \qquad B_{lm,js} := \beta_l(k) \, \delta_{lj} \delta_{ms}.$$

The linear equation (4.13) for the coefficients is then equivalent to

$$\sum_{j=0}^{\Lambda} \sum_{s=-j}^{j} (A - B\mathcal{M})_{lm,js} \, b_{js} = 0.$$

A non trivial solution exists, if the determinant $\det(A - B\mathcal{M})$ vanishes. The matrices A, B are connected to the definition of the scattering phase in (4.7)

$$(A + iB)_{lm,js} = \alpha_l(k)\, \delta_{lj}\delta_{ms} + i\beta_l(k)\, \delta_{lj}\delta_{ms}$$
$$= (\alpha_l(k) + i\beta_l(k))\, \delta_{lj}\delta_{ms}$$

Analogously

$$(A - iB)_{lm,js} = (\alpha_l(k) - i\beta_l(k))\, \delta_{lj}\delta_{ms}$$
$$\Rightarrow (A - iB)^{-1}_{lm,js} = \frac{1}{\alpha_l(k) - i\beta_l(k)}\, \delta_{lj}\delta_{ms}$$

The scattering phase, expressed in the matrix elements of A, B is then

$$e^{2i\delta_l} = \frac{(A + iB)_{lm,js}}{(A - iB)_{lm,js}}.$$

Using the scattering phase, one can rewrite the condition for the determinant

$$\det\left(e^{2i\delta} - U\right) \stackrel{!}{=} 0, \qquad U = \frac{\mathcal{M} + i}{\mathcal{M} - i}.$$

The zeros of the determinant determines k and thus the energy eigenvalues.

Again, the special case for $l = 0$ and $m = 0$ simplifies the above equation. In particular, the above matrix equations involving A and B turn into simple scalar equations. The relation for the scattering phase is then

$$e^{2i\delta_0} = \frac{\alpha_0 + i\beta_0}{\alpha_0 - i\beta_0}.$$

The energy eigenvalues k are then determined by

$$e^{2i\delta_0(k)} = \frac{\mathcal{M}(k) + i}{\mathcal{M}(k) - i}$$

$\mathcal{M}$ is just a scalar and is specified in the following steps.

4. **Energy spectrum in the A_1^+ sector**

So far, the energy eigenvalues k were considered in a finite box. In the following the energy spectrum will be discussed in a discretized box with cubic symmetry. The Hamiltonian is invariant under discrete rotations. Hence, its eigenstates can be characterized in irreducible representations of the cubic group. The representations of the cubic group are derived from the irreducible representations of $O(3)$. The spherical harmonics $\mathcal{Y}_{lm}$ are a basis in the space of all harmonic polynomials of degree l. Their transformation under a rotation $R \in O(3)$ is given by

$$\mathcal{Y}_{lm}(R\vec{r}) = \sum_{s=-l}^{l} D_{ms}^{(l)}(R) \, \mathcal{Y}_{ls}(\vec{r}). \tag{4.14}$$

The representations of the full cubic group $O(3, \mathbb{Z})$ is given by the representations of the special cubic group and parity. The parity of the harmonic polynomials is $P = -1^l$. The A_1^+ sector is then obtained for $l = 0$. The $(+)$ sign indicates the positive parity. A_1^+ corresponds to the trivial representation and thus $D_{ms}^{(l)}(R)$ is just a scalar. In this work, only the A_1^+ sector will play a role and the further discussion is restricted to this sector.

The condition for the determinant is then

$$e^{2i\delta_0} = \frac{\mathcal{M} + i}{\mathcal{M} - i}. \tag{4.15}$$

where $\mathcal{M}$ is just a scalar in this case. Using equations (4.10) and (4.11)

$$\mathcal{M} = \frac{1}{\pi^{\frac{3}{2}} q} \mathcal{Z}_{00}(1; q^2), \qquad q = \frac{kL}{2\pi}.$$

5. **The scattering phase and the two particle energy**

The last equation for the determinant (4.15) can be split up in its real part and its imaginary part. A simple calculation for the real part then yields

$$\tan(\delta_0) = \frac{1}{\mathcal{M}}.$$

The final result is then given by

$$\tan \delta_0(k) = \frac{\pi^{\frac{3}{2}} q}{\mathcal{Z}_{00}(1; q^2)}, \qquad q = \frac{kL}{2\pi}. \tag{4.16}$$

The final result for the non relativistic case given in equation (4.16) can now be transferred into quantum field theory by using the arguments at the beginning. The momentum variable k is then not any more given by the quantum mechanical energy eigenvalue $E = \frac{k^2}{2\mu}$ but by the relativistic formula

$$W = 2\sqrt{m^2 + k^2}. \tag{4.17}$$

The mass m denotes the mass of the particle in the initial state. It was assumed that the initial state particles are spinless and have equal masses. The restriction suites to the situation considered here. The method is more general and is able to incorporate distinct masses and spins. Furthermore, the arguments were restricted in order to analyse resonances even though it could equally well be used to analyse bound states which is then associated to a negative energy of the two particle system.

The above result gives the connection between the two particle energy eigenvalues and the scattering phase. The energy levels can be computed from lattice simulations and hence, the momentum k is given by inverting equation 4.17. k is then an arbitrary number and is not restricted to the lattice momentum. The arguments at the beginning in fact excluded the case where k takes values which coincide with some lattice momentum. This restriction is not necessary and the general derivation in the deferred work treats these momenta separately. Once the momentum k is known, the scattering phases are accessed by solving the final equation (4.16) numerically. From the discussion at the beginning of this chapter one knows that the total cross section is associated with the scattering phase (see (4.4)) and near a resonance it closely resembles a Breit-Wigner curve. The Breit Wigner curve is parametrized by the resonance mass and its width and thus a fit to the numerically obtained scattering phases yields the resonance parameters. In the above argument it was assumed that the only contribution to the A_1^+ sector arises from the $l = 0$ angular momentum configuration. In fact this is not true. The next angular momentum contribution comes from $l = 4$. However, at low momentum k it is known that the phase shifts $\delta_l(k)$ are suppressed by $\delta_l(k) \propto k^{2l+1}$ [52]. In the following it will be assumed that the lowest angular momentum contribution dominates and all higher momenta can be neglected.

The above result is valid for the centre of mass frame. In order to compute the two particle energies from lattice simulations, one has to consider particles with opposite spatial momenta. Due to the fact that the lattice momenta are restricted to the set given in (4.9) the total centre of mass energy gets easily beyond the inelastic threshold where the above result is not valid any more.

The analysis of scattering phases has been extended to moving frames in [54] where one of the two particles is at rest. This method allows to compute scattering phases for smaller energy levels and the method is complementary to the centre of mass frame and allows to compute more data for the scattering phases from the same configurations. The obtained two particle energies have to be translated back to the centre of mass frame. In general, the choice of a moving frame implies that one has to consider irreducible representations of a sub group of the cubic group. The remaining symmetry depends on the selected directions of the moving frame. The corresponding irreducible representations are given in [54]. The modification of the relation between the two particle energy in the moving frame and the scattering phase is given by

$$\tan \delta_0(q) = \frac{\gamma q \pi^{\frac{3}{2}}}{\mathcal{Z}^d_{00}(1; q^2)}, \qquad q = \frac{p^* L}{2\pi}.$$

p^* denotes the momentum which has been transferred back to the centre of mass frame by a Lorentz boost. The modified Zeta function is defined by

$$\mathcal{Z}^d_{00}(s; q^2) = \frac{1}{\sqrt{4\pi}} \sum_{r \in P_d} \frac{1}{(r^2 - q^2)^2}, \qquad P_d := \left\{ \vec{r} \in \mathbb{R}^3 | \vec{\gamma}^{-1} \left(\vec{n} + \frac{1}{2}\vec{d} \right), \vec{n} \in \mathbb{Z}^3 \right\}.$$

The vector $\vec{d}$ is related to the total momentum of the moving frame $\vec{P}$. γ is the usual Lorentz factor. Below are some definitions which are needed to compute the modified Zeta function

$$\vec{P} := \frac{2\pi}{L} \vec{d}$$

$$\gamma = \frac{1}{\sqrt{1 - v^2}}, \qquad \vec{v} = \frac{\vec{P}}{W_L}$$

$$\vec{\gamma}^{-1} \vec{n} = \gamma^{-1} \vec{n}_\parallel + \vec{n}_\perp.$$

W_L denotes the two particle energy in the moving frame which can be computed with time slice correlators. The corresponding observables are defined in the next section. $\vec{n}_\parallel$ and $\vec{n}_\perp$ is a decomposition of the vector $\vec{n}$ in its parallel and perpendicular parts with respect to the centre of mass velocity $\vec{v}$

$$\vec{n}_\parallel := \frac{(\vec{n} \cdot \vec{v}) \vec{v}}{v^2}$$

$$\vec{n}_\perp := \vec{n} - \vec{n}_\parallel.$$

4.2.2 Numerical results

As mentioned in the last section, the Higgs boson decays dominantly to any even number of Goldstone bosons, if kinematically allowed. The physical set-up chosen here allows

4.2 Resonance mass and width of the Higgs boson

always for such a decay. The bare parameters as well as the physical cut off, the Higgs boson propagator mass, the Goldstone boson mass and the obtained Top quark mass are summarized in table 4.1. Details on the finite size analysis and the extrapolation to infinite volume are described below.

Table 4.1: The table summarizes the bare parameters for the Monte Carlo simulations which were performed in order to determine the scattering phases. The next columns show the Higgs boson mass extracted from the propagator, the Goldstone boson mass quark mass and the renormalized *vev* . The last column shows the cut off (Λ). The latter physical quantities are obtained after an extrapolation to infinite volume. The large statistical uncertainty at $\lambda = 1.0$ for M_H^p is owed to some technical difficulties at the computing centre. Only lattice volumes up to $20^3 \times 40$ could be considered for the final extrapolation of the Higgs boson propagator to infinite volume.

κ	$\hat{\lambda}$	$\hat{y}$	J	M_H^p	M_G^p	m_t [GeV]	v_R	Λ [GeV]
0.12950	0.01	0.36274	0.001	0.278(1)	0.085(2)	174(1)	0.2786(3)	883(1)
0.24450	1.0	0.49798	0.002	0.386(28)	0.133(4)	179(2)	0.1637(5)	1503(5)
0.30200	∞	0.57390	0.002	0.405(4)	0.129(1)	178(1)	0.1539(2)	1598(2)

The Goldstone theorem ensures that the Goldstone bosons are massless. Due to an external current which couples to one of the scalar fields in the complex $SU_W(2)$ doublet, the symmetry is broken explicitly in the Lagrangian. The Goldstone bosons acquire a mass and they form a vector under cubic rotations. The magnitude of the current J is chosen such that the ratio of the Higgs boson mass M_H^p to the Goldstone boson mass is roughly 3. Here and below the superscript p in M_H^p and M_G^p denotes that the mass was extracted from the analysis of the momentum space propagator and a fit formula motivated from perturbation theory. The resonance mass which corresponds to the physical Higgs boson mass is obtained with the help of the correlation matrix analysis [49, 11] and a fit of the corresponding scattering phases to the generic Breit-Wigner curve.

Figure 4.4 shows the infinite volume extrapolation of the renormalized scalar *vev* and the top quark mass obtained from the fermion time slice correlator. The infinite volume results are obtained after a linear fit to the data starting from lattice volumes of at least

$16^3 \times 40$. The procedure above reflects the method which has been followed in order to determine the mass bounds of the Higgs boson [31, 30].

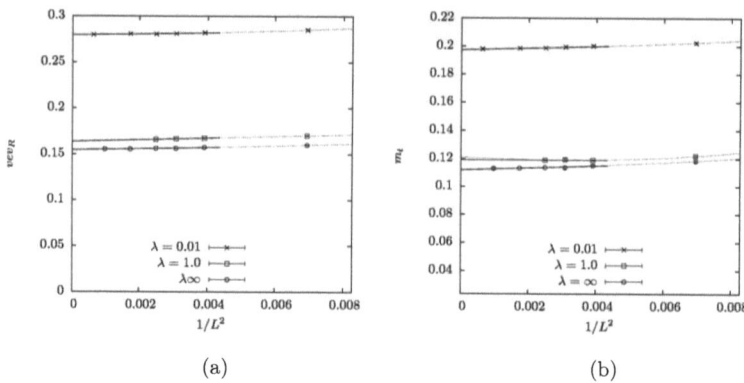

(a)   (b)

Figure 4.4: The figure shows the finite size effects of the scalar vev and the fermion mass for the three selected values of the quartic coupling. The top quark mass is computed from the fermion time slice correlation function. The figure shows also an extrapolation to infinite volume starting from lattice volumes of at least 16^3.

In the following, the results on the scattering phases and the correponding cross sections are presented. The scattering phases are computed from the energy levels which are obtained from the analysis of the correlation matrix given below. Once the scattering phases are computed, the optical theorem provides the connection to the total cross section. Near the resonance, the cross section exhibits the form of a relativistic Breit-Wigner function which will be used as a fit function in order to extract the resonance mass and the width. Finally, an infinite volume extrapolation of the energy eigenvalues are presented and contrasted to the Higgs boson resonance mass.

The Higgs field is a singlet under cubic rotations and transforms as elements in the A_1^+ representation. The two particle energies discussed in the previous section are constructed from the two particle Goldstone singlet $\mathcal{G}^T\mathcal{G}$ as it has the same quantum numbers as the Higgs boson.

The analysis of the resonance parameters involves several lattice volumes with identical bare parameters in order to compute the momentum dependence of the scattering phase.

4.2 Resonance mass and width of the Higgs boson

As shown in table 4.1, there are three distinct set of simulation parameters which shall be characterized with the value of the bare quartic coupling ($\lambda \in \{0.01, 1.0, \infty\}$). For each of the three values of the quartic coupling the simulations were performed on lattice volumes up to 40^4. The temporal extent is always set to 40. Table 4.2 shows the lattice volumes for the three different quartic couplings.

Table 4.2: The following table lists the spatial extent L_s of the lattice volumes $L_s^3 \times L_t$. The temporal extent is always set to 40. Furthermore, the approximate auto correlation times computed according to [57] is given in the last column.

λ	L_s	τ
$\lambda = 0.01$	12, 16, 18, 20, 24, 32, 36, 40	< 2.0
$\lambda = 1.0$	12, 16, 18, 20, 24, 36, 40	≈ 5
$\lambda = \infty$	12, 16, 18, 20, 24, 32, 40	< 3

In the following the two particle energies of the two Goldstone boson states will be discussed. Once these energy levels are known, the unstable nature of the Higgs boson can be studied by the method described at the beginning of this chapter.

The Goldstone bosons are stable particles such that their ground state energy can be calculated from the two point time correlation function. The concept of time correlators is widely used in lattice field theory and there are reliable techniques to extract mass eigenvalues from such correlators. The method of choice in this work, is the analysis of the correlation matrix [11]. The correlation matrix in the centre of mass is defined by

$$C_{\alpha\beta}^{cm}(\Delta t) := \frac{1}{L_t} \sum_{|t-t'|=\Delta t} \langle O_\alpha(t) O_\beta(t') \rangle_c .$$

where L_t is the tomporal size of the lattice. Throughout this chapter the temporal extent will be $L_t = 40$. The subscript c denotes that the disconnected part of the correlator has been subtracted. It has been shown in [49] that the eigenvalues of the correlation matrix decay exponentially with rising time separation Δt. The strength of the exponential decay is determined by the energy levels of the particle states, which are interpolated by the corresponding operators. The advantage of this method in contrast to the correlation function, which also exhibits an exponential decay, is that the energy eigenvalues computed from the correlation matrix respect the mutual interaction of different operators $(O_\alpha(t)O_\beta(t'), \alpha \neq \beta)$ which are in general complex such that they cannot be accessed in a straight forward manner from the time correlation function. Furthermore, the corrections to the energy levels obtained from the correlation matrix analysis is of order $\mathcal{O}\left(e^{-\Delta E_{N+1,n}t}\right)$. N denotes the number of independent observables considered in the correlation matrix. $\Delta E_{N+1,n}$ is the difference of energy levels $\Delta E_{N+1,n} = E_{N+1} - E_n$ [11]. In the following the operators which contribute to the two Goldstone system are collected.

The definition of the observables in the centre of mass frame are straight forward

$$\mathcal{O}_0(t) := \tilde{H}(\vec{0}, t)$$

$$\mathcal{O}_1(t) := \frac{1}{\sqrt{3}} \frac{1}{|Q_1|} \sum_{\vec{n} \in Q_1} \tilde{\mathcal{G}}^T(\vec{n}, t)\tilde{\mathcal{G}}(-\vec{n}, t)$$

$$Q_1 = \left\{\vec{n} \in \mathbb{Z}^3 | n^2 = 0\right\}, \qquad |Q_1| = 1$$

$$\mathcal{O}_2(t) := \frac{1}{\sqrt{3}} \frac{1}{|Q_2|} \sum_{\vec{n} \in Q_2} \tilde{\mathcal{G}}^T(\vec{n}, t)\tilde{\mathcal{G}}(-\vec{n}, t)$$

$$Q_2 = \left\{\vec{n} \in \mathbb{Z}^3 | n^2 = 1\right\}, \qquad |Q_2| = 6$$

$$\mathcal{O}_3(t) := \frac{1}{\sqrt{3}} \frac{1}{|Q_3|} \sum_{\vec{n} \in Q_3} \tilde{\mathcal{G}}^T(\vec{n}, t)\tilde{\mathcal{G}}(-\vec{n}, t)$$

$$Q_3 = \left\{\vec{n} \in \mathbb{Z}^3 | n^2 = 2\right\}, \qquad |Q_3| = 12.$$

The correlation matrix is thus a 4×4 matrix.

In order to collect more data on the scattering phases, the modification of the method to a moving frame was analysed as well. The moving frame is characterized by a constant vector $\vec{d}$ which indicates the momentum of the frame. The observables for the moving frame are constructed such that one of the Goldstone bosons is at rest while the other can take any momentum allowed on the lattice. The selection of a constant vector $\vec{d}$ breaks the cubic symmetry and thus special care is needed while constructing the observables. Fortunately, it turns out that the A_1^+ sector does not need much modification and explicit relations are given in [54]. The lowest energy eigenstates are associated to the lowest possible relative momentum and thus only moving frames with momentum $\vec{d} = (0, 0, 1)$

4.2 Resonance mass and width of the Higgs boson

and permutations thereof will be considered. The observables are

$$\vec{d}_i = \vec{e}_i$$
$$\mathcal{O}_{\vec{d}_i,0}(t) := \tilde{H}(\vec{d},t)$$
$$\mathcal{O}_{\vec{d}_i,1}(t) := \tilde{\mathcal{G}}^T(\vec{d}_i,t)\tilde{\mathcal{G}}(\vec{0},t)$$
$$\mathcal{O}_{\vec{d}_i,2}(\Delta t) := \frac{1}{4} \sum_{\vec{n} \in Q_{d_i,2}} \tilde{\mathcal{G}}^T(\vec{n}+\vec{d}_i,t)\tilde{\mathcal{G}}(-\vec{n},t)$$
$$Q_{d_i,2} = \left\{\vec{n} \in \mathbb{Z} | \vec{n} \cdot \vec{d}_i = 0, n^2 = 1\right\}$$
$$\mathcal{O}_{\vec{d}_i,3}(t) := \tilde{\mathcal{G}}^T(2\vec{d}_i,t)\tilde{\mathcal{G}}(-\vec{d}_i,t).$$

$\vec{e}_i$ is the unit three-vector in direction i. The correlation matrix in the moving frame is then given by

$$C^{mf}_{\alpha\beta}(\Delta t) := \frac{1}{3}\frac{1}{L_t} \sum_{|t-t'|=\Delta t} \sum_{i=1}^{3} \left\langle O^*_{\vec{d}_i,\alpha}(t)\, O_{\vec{d}_i,\beta}(t') \right\rangle_c$$
$$\alpha,\beta \in \{1,2,3\}\,.$$

The energy levels obtained from the moving frame are connected to energy levels in the corresponding centre of mass frame by Lorentz transformation. The collection of all energy levels in both frames are presented in table C2 and C3 in appendix C.

Figure 4.5 shows the energy eigenvalues from the correlation matrix analysis. Only those eigenvalues are displayed which belong to the ground state of the Higgs boson. The figure also shows an extrapolation to infinite volume by a linear fit. All obtained mass eigenvalues are in good agreement to the fitted line and no higher order corrections in $\frac{1}{L_s^2}$ are visible. Due to the smaller errors for the smaller lattice volumes, the linear fit is dominated by the latter. Therefore, it is sufficient to consider only lattice volumes up to $32^3 \times L_t$ in order to perform an extrapolation of the generalized eigenvalues to infinite volume. Here however, all available lattice volumes were considered.

The relative momentum k in the rest frame is given by inverting the equation

$$W_k = 2\sqrt{m_G^2 + k^2} \tag{4.18}$$

The Goldstone boson mass is obtained by a separate correlation matrix analysis involving the three one particle Goldstone fields.

The two particle energies in the moving frame have to be transformed to the centre

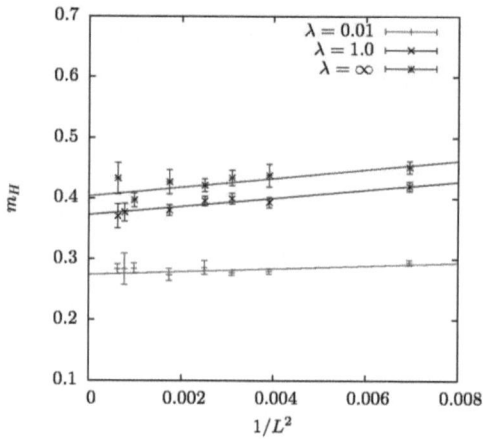

Figure 4.5: The figure shows the energy eigenvalues of the Higgs boson obtained by the correlation matrix analysis. The eigenvalues are plotted against the inverse squared lattice size in order to perform an extrapolation to infinite volume.

of mass frame by Lorentz transformation. Once the centre of mass energy is known, the relative momentum is given again by inverting equation (4.18).

Figure 4.6 shows the obtained scattering phases for the three different physical situations. The scattering phase takes values in the intervall $[0, \pi]$ and is plotted against the momentum k. If the scattering phase $\delta(k)$ passes through $\frac{\pi}{2}$ it indicates the existence of a resonance. Hence, all three set-ups involve an unstable Higgs boson and its resonance parameters are obtained by a fit of the obtained scattering phases to the Breit-Wigner function. The cross section can be decomposed into spherical harmonics and is then given by

$$\sigma(k) = \frac{4\pi}{k^2} \sum_{j=0}^{\infty} (2j+1) \sin^2\left(\delta_j(k)\right) \qquad (4.19)$$

$$\approx \frac{4\pi}{k^2} \sin^2\left(\delta_0(k)\right). \qquad (4.20)$$

As was mentioned in the beginning of this chapter and in Chapter 2, the total cross section resembles a Breit-Wigner curve near a resonance. As argued before, the contribution of the higher angular momenta $j > 0$ are neglected. Here the Breit Wigner function is used as a fit function in order to extract the resonance mass and the width. The first column

4.2 Resonance mass and width of the Higgs boson

in figure 4.6 shows the cross sections and the Breit-Wigner fit. The explicit form of the fit function is

$$f(k) := 16\pi \frac{M_H^2 \Gamma_H^2}{(M_H^2 - 4m_G^2)\left((W_k^2 - M_H^2)^2 + M_H^2 \Gamma_H^2\right)}.$$

The solid curve in the second column in figure 4.6 is then obtained by inverting equation (4.19) which gives the scattering phases.

Finally table 4.3 summarizes the results obtained by the different approaches. The physical Higgs boson mass is compared to the mass obtained from the Higgs propagator and the energy eigenvalues obtained with the help of the correlation matrix analysis. The latter results were obtained after an extrapolation to infinite volume.

Table 4.3: The table summarizes the obtained final results on the resonance mass and the resonance width of the Higgs boson. $\hat{\lambda}$ denotes the bare quartic coupling. The first line is a preliminary result from Chapter 8 in [25]. Λ is the cut off of the theory. The following two columns display the resonance parameters computed from the scattering phases. Γ_H^p is the width obtained from perturbation theory where a non vanishing mass for the Goldstone bosons has been considered. Finally the mass extracted from the propagator as well as the mass eigenvalues computed with the help of the correlation matrix is shown. The latter results were obtained after an extrapolation to infinite volume as shown in figure 4.4 and figure 4.5.

$\hat{\lambda}$	Λ [GEV]	Res. mass M_H	Res. width Γ_H	Γ_H^p	Prop. mass M_H^p	GEVP
0.01	593(1)	0.428(3)	0.009(3)	0.0076(2)	0.433(3)	
0.01	883(1)	0.2811(6)	0.007(1)	0.0054(1)	0.278(2)	0.274(4)
1.0	1503(5)	0.374(4)	0.033(4)	0.036(8)	0.386(28)	0.372(4)
∞	1598(2)	0.411(3)	0.040(4)	0.052(2)	0.405(4)	0.403(7)

The analysis of the resonance parameters of the Higgs boson within the pure Higgs-Yukawa model shows that at a cut off of about 1.5 TeV, the width is at most 10% of the resonance mass. Figure 4.7 shows the obtained resonance widths of the Higgs boson against the renormalized quartic coupling. The figure also shows the width expected from perturbation theory. Furthermore, the resonance mass is in perfect agreement with the mass obtained from the propagator as well as the Higgs mass extracted from the generalized eigenvalue problem. The simulations at $\lambda = 1.0$ and $\lambda = \infty$ belong to a cut off of around 1.5 TeV. It was necessary to reduce the cut off to 880 GeV for the smallest quartic coupling $\lambda = 0.01$ in order to meet the resonance condition ($\frac{M_H^p}{m_G} \approx 3$). As one can see from table 4.3, the Higgs mass is well below the cut off. Especially for the smallest quartic coupling $\lambda = 0.01$ the resonance region is very small ($m_G = 0.09(1) \Rightarrow 0.18 \leq W_k \leq 0.36$) which in turn necessitates large lattice volumes in order to obtain energy eigenvalues which lead to scattering phases near the resonance mass. The plots in figure 4.6 show that the analysis of the moving frame is of great importance to extract reliable results.

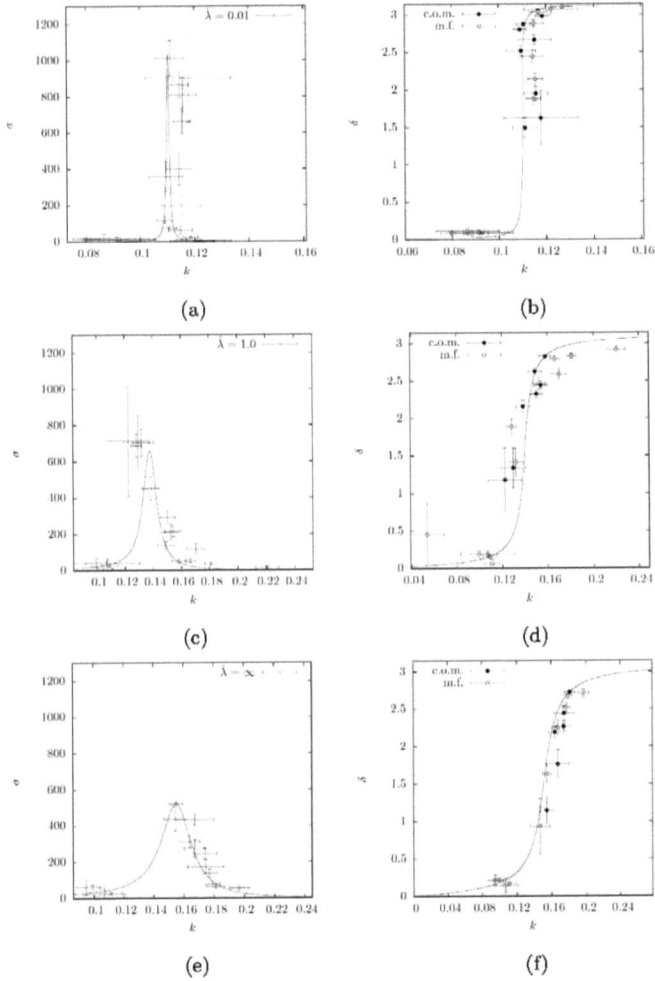

Figure 4.6: The figure shows the scattering phases obtained in the three different physical situations for $\lambda \in \{0.01, 1.0, \infty\}$ ordered vertically. The filled circles refer to scattering phases obtained from the analysis in the centre of mass frame as originally proposed in [46]. The empty citcles denote the scattering phases computed within a moving frame. The modification was proposed in [54]. The vertical dotted line indicates the inelastic threshold. The computations were performed on various lattice volumes $L_s^3 \times 40$ where $L_s \in \{12, 16, 18, 20, 24, 32, 40\}$.

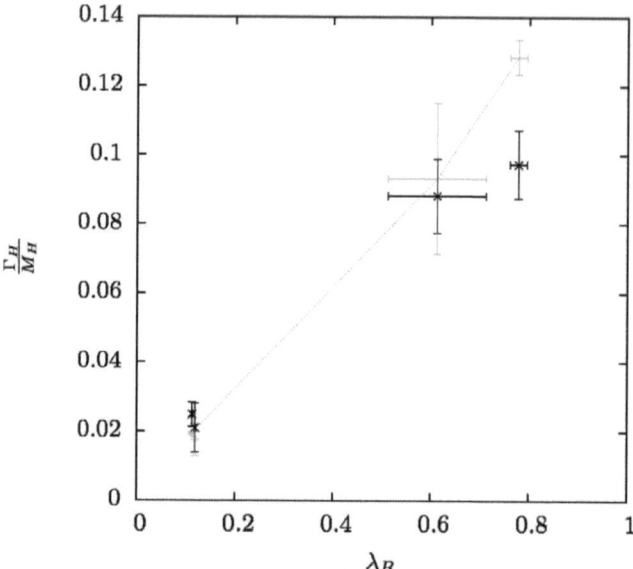

Figure 4.7: The figure summarizes the obtained values for the resonance width of the Higgs boson at various quartic couplings. The X-Axis denotes the renormalized quartic coupling which is defined by $\lambda_R = \frac{M_H^{p\,2} - M_G^{p\,2}}{8v_R^2}$. The Y-Axis shows the relative width of the Higgs boson. The grey points denote the Higgs boson resonance width computed within perturbation theory where the Goldstone bosons were considered as generic massive scalar particles. The dashed line connects the points obtained from perturbation theory.

5 Beyond the Standard model: A fourth generation of fermions

Though the richness and precision of theoretical predictions in the framework of the standard model is unique and remarkable, its well known deficits necessitates to think about extensions of it. Even ignoring the main conceptional caveat, that gravity is not included at all and cannot be included in a straight forward way, the model fails to describe phenomena which are accessible in present experiments and observations. The most prominent of those are

- the baryon asymmetry in the universe,
- the lack of a candidate for dark matter and dark energy,
- the strong CP problem.

An extension of the standard model with a fourth generation of heavy quarks and leptons (SM4) arranged within a $SU_L(2)$ doublet permits to alter the model in a way such that it is compatible with electroweak precision measurements. A fourth generation of fermions provides various prospects to augment the model to enable a deeper understanding of flavour physics and mass hierarchies [40]. The main motivation, however, is that SM4 may satisfy the three Sakharov conditions [55] such that the observed Baryon asymmetry of the universe is consistent with theory [16, 41]. It is important to mention that the standard model (in the following SM3) does not allow for a strong first order phase transition and the only CP violating source in SM3 can be parameterised with a phase in the CKM matrix which is known to be too small to explain the Baryon asymmetry.

Another source of CP violation within SM3 is the term

$$\bar{\theta} G_{\mu\nu} \tilde{G}^{\mu\nu}$$

which can be added to the standard model Lagrangian as it is consistent with the symmetries of the model. $\bar{\theta}$ is a measure for CP violation induced by strong interactions. Experimental data, however, constraints this parameter to be smaller than 10^{-10}. At this point it shall be mentioned that there are other modifications of SM3 without involving further fermions, such as the axion mechanism (see [21] and references therein), which may provide enough CP violation in order to satisfy the Sakharov condition.

The main arguments against a fourth generation were given in the beginning of 1990 where precise measurements of the Z boson peak and width were performed in SLAC and CERN [3, 4, 5, 18, 7, 2]. Assuming that the neutrino mass of a potential fourth generation is below $M_Z/2 \approx 45$ GeV it could be proven that there are indeed just three generations [10]. An alternative method was followed by analysing the S and T parameter as described in [Frampton:1999xi]. Assuming that the heavy fermion doublet is degenerate it was concluded that three fermion generations are favoured. A *non-degenerate* fermion doublet however, reduces this tension and allows a further heavy fermion generation in the sense that the deviation from the measured value of the S parameter is below 3σ [21].

SM4 does not present a solution for all deficits within SM3 but it may be a viable extension towards a more complete theory. A detailed and comprehensive review on this topic is given in [21].

The existence of a fourth generation with its strong coupling to the scalar sector necessitates to reinvestigate the Higgs boson mass bounds. Moreover, the large Yukawa couplings may give rise to genuine non-perturbative effects which can be treated with the numerical method described in chapter 2.

In this chapter the Higgs boson mass bounds are evaluated in the presence of a fourth generation of heavy quarks. Similar to the argument in the case of the standard model, only the heaviest fermion doublet will be considered. Further details on the model is given in section 5.1. The focus of the computation is to determine the mass bounds for varying values of the cut off and a fixed heavy degenerate quark doublet with a mass of about 700 GeV. In a second set-up the heavy quark mass is varied between 200 GeV and 700 GeV while the cut off is held constant at about 1500 GeV.

The final results from the lattice simulations are obtained after an extrapolation to infinite volume. The main outcome is that the lower Higgs boson mass bound strongly depends on the mass of the heavy quarks while the upper Higgs boson mass is only slightly enhanced with respect to the standard model.

Although the simulation strategy and the relevant observables have been discussed in Chapter 2, some aspects are repeated in this chapter in order to give a self consistent and complete presentation.

5.1 The model

Heavy fermions are realized through large couplings to the scalar doublet. These large couplings dominate the contribution to the Higgs mass with respect to the weak gauge bosons and the light fermions. Thus it is reasonable to neglect the weak gauge bosons and restrict the model solely to the heavy fourth generation of fermions. Moreover, a degenerate doublet of heavy quarks is considered in order to get a first estimate of the Higgs boson mass bounds. The model is then given by the *Euclidean* Lagrangian

$$L_E^{HY} = \frac{1}{2}(\partial_\mu \varphi)^\dagger \cdot (\partial^\mu \varphi) + \frac{1}{2}m^2 \varphi^\dagger \cdot \varphi + \lambda \left(\varphi^\dagger \cdot \varphi\right)^2$$
$$+ \overline{t'} \slashed{D} t' + \overline{b'} \slashed{D} b' + y'_b \begin{pmatrix} \overline{t'} \\ \overline{b'} \end{pmatrix}^T_L \cdot \varphi \, b'_R + y'_t \begin{pmatrix} \overline{t'} \\ \overline{b'} \end{pmatrix}^T_L \cdot \tilde{\varphi} \, t'_R + h.c.. \quad (5.1)$$

$\tilde{\varphi}$ transforms like a $SU(2)$ vector and is given by

$$\tilde{\varphi} = i\tau_2 \varphi^*, \qquad \tau_2 \text{ is the Pauli matrix}.$$

The following analysis have been performed by means of Monte Carlo simulations. The method is based on a discretized space time lattice. The above Lagrangian exhibits a global $SU(2)_W \times U(1)_Y$ symmetry. Implementing chiral symmetry on such a lattice involves several difficulties. However, a consistent lattice modified chiral symmetry can be defined with the help of the Neuberger overlap operator [51]. The lattice modified chiral symmetry converges in the continuum limit to the desired continuum symmetry.

Details on the simulation algorithm and the lattice modified chiral symmetry are given in Chapter 2. The Euclidean lattice action is given by

$$S = -\kappa \sum_{x,\mu} \Phi_x^\dagger \left(\Phi_{x+\mu} + \Phi_{x-\mu}\right) + \sum_x \Phi_x^\dagger \Phi_x + \hat{\lambda} \sum_x \left(\Phi_x^\dagger \Phi_x - N_f\right)^2$$
$$+ \sum_{x,y} \overline{\psi}_x^\alpha \left\{ \mathbb{I}_2 D_{x,y}^{\alpha\beta} + \hat{y}' \left(P_- \phi \hat{P}_- + P_+ \phi^\dagger \hat{P}_+\right)_{x,y}^{\alpha\beta} \right\} \psi_y^\beta. \quad (5.2)$$

ψ is the spinor doublet of the heavy t' and b' quarks

$$\psi = \begin{pmatrix} t' \\ b' \end{pmatrix}. \quad (5.3)$$

The scalar fields are arranged in a quarternion

$$\phi := \begin{pmatrix} \tilde{\varphi}_1 & \varphi_1 \\ \tilde{\varphi}_2 & \varphi_2 \end{pmatrix} =: \phi^0 \mathbb{I} - i\sigma^j \phi^j, \quad \phi_\mu \in \mathbb{R}.$$

Furthermore, it is common to scale the fields with a factor of $\sqrt{2\kappa}$

$$\Phi^\mu := \frac{1}{\sqrt{2\kappa}} \phi^\mu.$$

The couplings in the Lagrangian (5.1) are then recovered with the relations

$$\lambda = \frac{\hat{\lambda}}{4\kappa^2}, \quad m_0^2 = \frac{1 - 2N_f\hat{\lambda} - 8\kappa}{\kappa}, \quad y_{t,b} = \frac{\hat{y}_{t,b}}{\sqrt{2\kappa}}.$$

Details on the simulation algorithm is given in [25].

The ϕ^4 theory is assumed to be trivial in space time dimensions larger than three. Aizenman proved analytically [6] that the model is indeed trivial in dimensions larger than five. There were many numerical studies which were consistent with the triviality picture also for four space time dimensions. The triviality is inherently connected to the scalar sector and thus also the Higgs-Yukawa model is assumed to be trivial in four space time dimensions. That means any set of bare parameters yields a free, i.e. gaussian theory as the cut off gets arbitrarily large. An interacting Higgs-Yukawa model therefore implies a finite intrinsic cut off and renormalized quantities implicitly depend on this cut off.

As the Higgs boson has not yet been discovered experimentally, its mass bounds at a given cut off are of great importance for phenomenology. The strategy in this work is to evaluate the model at the physical point of interest and to derive the upper and lower Higgs boson mass bounds which are attainable within this model.

The bare parameters of the theory are λ_0, m_0 (or, equivalently κ) and y_0. The subscript zero denotes that all considered parameters are not renormalized. The observables which are evaluated within this work and the addressed questions focuses on the broken phase of the model. The parameter κ respectively m_0 has thus to be chosen such that the simulation point is above the phase transition line. Furthermore the obtained non-zero magnetization, which indicates the broken phase, is set such that the scalar vev meets the phenomenologically known value of 246 GeV. The latter scale is then used to determine the cut off (Λ) of the theory. Within the broken phase, the parameter κ or m_0 is tuned to achieve the desired value of the cut off. The calculations presented in this chapter aim to generate quark masses higher than the standard model top quark. Hence, the bare Yukawa coupling has to be tuned such that the masses of the heavy fourth generation quark doublet lies above 175 GeV. Finally λ_0 remains to be fixed. It was shown in [31] and [30] that the upper Higgs boson mass is reached at infinite bare quartic coupling while a vanishing bare quartic coupling yields the lower mass bound, as expected from perturbation theory.

5.2 Current Mass bounds related with a fourth generation

While experimental data constrains the lower bound of a potential heavy fourth generation quark mass, theoretical calculations relying on partial wave expansion and unitarity can provide an upper bound above which perturbation theory fails. The upper bound is either saturated or it defines an energy at which non perturbative effects dominate over the leading order prediction from perturbation theory.

The current lower bound for the b' quark mass is established by the CDF collaboration [1]

$$m_{b'} \geq 338 \text{ GeV} .$$

The perturbative method to determine an upper bound on the heavy fermion and the Higgs boson mass relies on unitarity and partial wave expansion. The differential cross section is given by the squared modulus of the scattering amplitude which in turn can be calculated in a perturbative expansion of Feynman diagrams

$$\frac{\mathrm{d}\sigma(\alpha \to \beta)}{\mathrm{d}\Omega} = |f(\alpha \to \beta)|^2,$$

$$f(\alpha \to \beta) = -\frac{4\pi^2}{E}\sqrt{\frac{k'E'_1 E'_2 E_1 E_2}{k}} \underbrace{\mathcal{M}_{\beta\alpha}}_{\text{Feynman amplitude}}.$$

The partial wave analysis is an expansion of the Feynman amplitude in terms of spherical harmonics. Unitarity of the partial wave amplitudes then defines an upper bound on the involved renormalized parameters. This method was used in [43] where an upper bound for the Higgs boson mass could be derived

$$M_H^2 = \frac{4\pi\sqrt{2}}{G_F} \approx 1 \text{ TeV}.$$

It is important to notice that this analysis was performed by truncating the Feynman amplitude such that contributions of the order $\frac{1}{M_W^4}$ could be neglected. This procedure is justified as long as perturbation theory is viable. Higgs boson masses above a TeV are not excluded and would just indicate the break down of perturbation theory.

The above method of partial wave unitarity was also applied for heavy fermions by [15]. The upper bound for heavy fermions was constrained to be lower than

$$m_{t'} \leq 550 \text{ GeV}.$$

5.3 Observables and extraction of mass eigenvalues

The simulation algorithm produces a set of scalar fields, a so called ensemble, which is sampled with a distribution corresponding to the action given in equation (5.2). The expectation value of any observable depending on the scalar field is then given by an ensemble average. The vacuum expectation value of the scalar field (*vev*) will be computed in order to distinguish the broken phase from the symmetric phase. The bare *vev* is defined by

$$vev := \sqrt{2\kappa}\, |\overline{\Phi}| = \sqrt{2\kappa} \left(\sum_\alpha \overline{\Phi}_\alpha^2\right)^{\frac{1}{2}} \tag{5.4}$$

$$\overline{\Phi}^\alpha = \frac{1}{V}\sum_{x \in \mathbb{Z}_L^4} \Phi_x^\alpha. \tag{5.5}$$

The straight forward definition of the scalar vacuum expectation value $\langle \varphi \rangle$ is not invariant under the symmetries of the model and thus the ensemble average necessarily vanishes. The usual procedure in the context of the path integral formulation is to add an external current which couples to one of the scalar field components.

$$S[J] := S + J \sum_x \Phi_x^0.$$

$\sqrt{2\kappa}\,\Phi^0$ is then identified with the Higgs field. The current J breaks the symmetry explicitly and expectation values of observables are a functional of the external current J. The physical result is then taken in the twofold limit

$$\langle O \rangle = \lim_{J \to 0} \lim_{V \to \infty} \langle O[J] \rangle.$$

The above limiting procedure implies an enormous numerical task. It was shown in [37, 36] that the definition (5.5) of the scalar vev converges to the vev obtained after taking the limits $V \to \infty$ and $J \to 0$. The vev is taken according to the defintion (5.5) throughout this chapter.

In order to compute the renormalized vev_R one needs the renormalization factor Z of the scalar field. It is known that the Higgs boson field renormalization factor and the Goldstone field renormalization factor are very close to each other. Here the latter will be used in order to determine the renormalized vev_R. The action defined in (5.2) does not depend on the lattice spacing a and consequently the obtained lattice results are expressed in lattice units. The scalar vev is a dimensionfull quantity and therefore the renormalized vev_R expressed in lattice units is given by

$$\frac{v_R}{a} := \frac{1}{\sqrt{Z_G}} \frac{v}{a}.$$

The renormalized vev_R is known phenomenologically from W boson scattering and the width of the Z boson resonance

$$246 \text{ GeV} = \frac{v_R}{a}$$
$$\Rightarrow \Lambda := \frac{1}{a} = \frac{\sqrt{Z_G}\, 246 \text{ GeV}}{v}.$$

The lattice spacing is then determined such that the renormalized vev_R assumes the phenomenologically known value of 246 GeV. Once the lattice spacing is fixed the cut off can be defined as the inverse lattice spacing.

5.3.1 Extraction of mass eigenvalues

The gloabl $SU_W(2) \times U_Y(1)$ symmetry allows to transform the scalar doublet such that it takes the form

$$\varphi_x = \begin{pmatrix} \mathcal{G}_x^2 + i\mathcal{G}_x^1 \\ v + H_x - i\mathcal{G}_x^3 \end{pmatrix}.$$

The Higgs and Goldstone boson propagator are then given by a discrete Fourier transformation

$$\tilde{H}_{\hat{p}} := \sum_x e^{-i\hat{p}x} H_x,$$

$$\tilde{\mathcal{G}}_{\hat{p}} := \sum_x e^{-i\hat{p}x} \mathcal{G}_x,$$

$$\tilde{G}_H^{-1}\left(\hat{p}^2\right) = \left\langle \tilde{H}_{-\hat{p}} \tilde{H}_{\hat{p}} \right\rangle,$$

$$\tilde{G}_G^{-1}\left(\hat{p}^2\right) = \frac{1}{3}\sum_{i=1}^{3} \left\langle \tilde{\mathcal{G}}_{-\hat{p}}^i \tilde{\mathcal{G}}_{\hat{p}}^i \right\rangle.$$

The hat above the momentum variable indicates the discrete lattice momentum $\hat{p} \in \Gamma_{L,T}$

$$\Gamma_{L,T} := \left\{ p \in \mathbb{R}^4 | p_0 = \frac{2\pi}{T} n_0, \quad p_i = \frac{2\pi}{L} n_i, \right.$$

$$\left. n_0 \in \mathbb{Z} : 0 \leq n_0 < T, \quad n_i \in \mathbb{Z} : 0 \leq n_i < L \right\}.$$

The time slice correlator is a widely used method to extract low lying energy eigenvalues from lattice simulations. Its time dependence can be derived from first principles. The time slice correlator is defined by

$$C(\Delta t) := \sum_{\substack{t,t' \\ |t-t'|=\Delta t}} \langle O(t)O(t') \rangle_c$$

$$\langle O(t)O(t') \rangle_c := \langle O(t)O(t') \rangle - \langle O(t) \rangle \langle O(t') \rangle.$$

In order to clarify the peculiarities of the time slice correlator some details are given below.

$$C(\Delta t) = \sum_{\substack{t,t' \\ |t-t'|=\Delta t}} \sum_n \langle \Omega|O(t)|n \rangle \langle n|O(t')|\Omega \rangle$$

5.3 Observables and extraction of mass eigenvalues

$|n\rangle$ are eigenstates of the Hamilton operator and fulfil the completeness relations

$$\sum_n |n\rangle\langle n| = \mathbb{I}$$

$$\Rightarrow C(\Delta t) = \sum_{\substack{t,t' \\ |t-t'|=\Delta t}} \sum_n \langle\Omega|e^{iHt}\, O(0)\, e^{-iHt}|n\rangle \langle n|e^{iHt'}\, O(0)\, e^{-iHt'}|\Omega\rangle$$

$$= \sum_{\substack{t,t' \\ |t-t'|=\Delta t}} \sum_n \langle\Omega|O(0)\, e^{-iE_n t}|n\rangle \langle n|e^{iE_n t'}\, O(0)|\Omega\rangle$$

$$= \sum_{\substack{t,t' \\ |t-t'|=\Delta t}} \sum_n e^{-iE_n \Delta t} \langle\Omega|O(0)|n\rangle \langle n|O(0)|\Omega\rangle$$

$$= \sum_{\substack{t,t' \\ |t-t'|=\Delta t}} \sum_n e^{-iE_n \Delta t} \left|\langle\Omega|O(0)|n\rangle\right|^2$$

The time slice correlator falls off exponentially and for large time separations $\Delta t \gg 1$ the lowest energy eigenvalue will dominate the exponential decay. Higher energy eigenvalues are only accessible if one can find suitable observables $O(t)$ such that their projection onto lower lying energy states $O(t)|n\rangle$ vanish. This is possible if these eigen states are separated from the lower eigenstates by distinct quantum numbers.

The case of the Higgs boson is more complicated. A one particle Higgs boson state has the same quantum numbers as a two Goldstone system. Due to the finite volume, the Goldstone particles acquire a mass. In the case were the bare quartic coupling is small and the one particle Higgs boson state has a lower energy eigenvalue than the two particle Goldstone system, the Higgs boson is a stable particle. Its energy eigenvalue can be determined reliably with the help of the time slice correlator. The upper mass bound, which is obtained at *infinite* bare quartic coupling, cannot be extracted with the correlator. The spectrum is dominated by the two particle Goldstone states and various excitations thereof. Another approach to extract the mass of the Higgs boson is to locate the pole of the propagator.

By definition, particle masses of fundamental particles are given by the pole of their propagator in momentum space. The Euclidean propagator of a scalar particle is given by

$$G_E^{-1}(p^2) = p^2 + m_0^2 - \Sigma(p^2; \Lambda).$$

In case that the particle of interest is a resonance, special care is needed. An unstable particle can decay into lighter particles which is reflected by a branch cut along the negative p^2 axis. The branch cut starts from the threshold value given by the two particle energies of the lighter particles in the theory. The pole of the analytic propagator is complex. In this case the renormalized mass is defined by the vanishing real part of the inverse Euclidean propagator

$$\Re\left\{G_E^{-1}\left(\bar{p}^2 = -m_R^2\right)\right\} = 0$$

$$\bar{p} = (im_R, 0, 0, 0).$$

The above definition is consistent with the relations given in [48].

The Euclidean propagator can be computed from lattice simulations but as the above equation shows, one needs an analytic continuation to negative squared momenta in order to locate the pole. This can only be achieved if an analytic expression for the Euclidean propagator is available. In this work, the functional structure motivated by one loop perturbation theory will be employed in order to locate the pole of the Higgs boson propagator. The one loop result is computed in detail in Chapter 3. The final result is

$$\left(G_H^R\left(p^2, M_H^2, M_G^2, \lambda_R, y_R; \Lambda\right)\right)^{-1} =$$
$$\frac{1}{Z}\left\{p^2 + M_H^2 + 18\left(4\lambda_R v\right)^2 \left(I_1(p^2, M_H^2, \Lambda) - I_1(-M_H^2, M_H^2, \Lambda)\right)\right.$$
$$+ 6\left(4\lambda_R v\right)^2 \left(I_1(p^2, M_G^2, \Lambda) - I_1(-M_H^2, M_G^2, \Lambda)\right)$$
$$\left. - \Delta\Sigma_H^f\left(p^2, M_f = y_r v_R, y_R; \Lambda\right)\right\}.$$

$$I_1(p^2, m^2, \Lambda) = \frac{1}{(4\pi)^2}\left\{1 + \ln\left(\frac{\Lambda^2}{m^2}\right)\right.$$
$$\left. - \sqrt{1 + \frac{4m^2}{p^2}} \ln\left(\frac{1 + \sqrt{1 + \frac{4m^2}{p^2}}}{-1 + \sqrt{1 + \frac{4m^2}{p^2}}}\right)\right\}.$$

Σ_H^f denotes the contribution of the fermions to the self energy of the Higgs boson. In all cases considered here, the branch cut induced by the fermions in the analytic continuation of the Higgs boson propagator lies beyond the branch cut induced by the Goldstone bosons. Due to statistical errors of the numerical data, it will not be possible to resolve the structure of the branch cut induced by the fermions. The functional structure is then reduced to

$$f_H\left(p^2, M_H^2, M_G^2, A, B, Z\right) =$$
$$\frac{1}{Z}\left\{p^2 + M_H^2 + A\left(I_1(p^2, M_H^2, \Lambda) - I_1(-M_H^2, M_H^2, \Lambda)\right)\right.$$
$$\left. + B\left(I_1(p^2, M_G^2, \Lambda) - I_1(-M_H^2, M_G^2, \Lambda)\right)\right\}.$$

In this work the upper as well as the lower Higgs boson mass bound will be determined on the basis of the Higgs boson propagator. In Chapter 3 a perturbative expansion of the Higgs boson propagator is given up to one loop and includes bosonic as well as fermionic loops. It is assumed that the functional structure of the one loop Higgs boson propagator is also valid for larger values of the renormalized quartic coupling. The perturbative expansion of the propagator will be used as a fit function in order to determine the parameters of f_H. After a successful fitting procedure, f_H can be continued into the complex p^2 plane and the pole can be located numerically for negative squared momenta.

Finally the fermion time slice correlator will be discussed. The fermion mass can be extracted from either the left or the right handed components of the spinor. In the following, only the left handed components are discussed, but it is straight forward to apply the arguments for the right handed components. The left handed correlator $C_f(t_0 - t_1)$ is given by

$$C_f(t_0 - t_1) := \left\langle \Re \operatorname{Tr} \left\{ \left(\hat{P}_L \psi\right)_{(t_0, \vec{p}=0)} \left(\bar{\psi} P_L\right)_{(t_1, \vec{p}=0)} \right\} \right\rangle.$$

The fermionic degrees of freedom are not directly accessible but the above expectation value can be reconstructed from the matrix elements of the fermion matrix as described at the end of Chapter 2.

5.4 Cut off dependent Higgs boson mass bounds

The Higgs boson mass bounds for the standard model top and bottom quark have been established in [31, 30]. This chapter aims to explore the effect on the mass bounds in the presence of a heavy degenerate top bottom doublet.

The perturbative upper bound for a heavy degenerate quark mass is assumed to be around 550 GeV. The simulations performed here therefore attempt to attain fermion masses above the perturbative unitarity bound. The following section shows the dependence of the Higgs boson mass at cut off values ranging from 1500 GeV to 3500 GeV while the heavy top- bottom quark mass was held fixed at around 676 ± 22 GeV. The particle masses and the vacuum expectation value of the scalar field can have strong dependence on the lattice size and therefore a finite size analysis and an extrapolation to infinite volume is inevitable.

Here the final results on the lower and upper Higgs boson mass bound are presented. The results are compared with the established results for the standard model fermions to visualize the effect of a fourth generation of fermions.

5.4.1 Numerical results

In order to vary the cut off, several simulations with different values of m_0 and y_0 had been performed. λ_0 is set to zero for the lower Higgs boson mass bound while the simulations for the upper Higgs boson mass bound are performed at infinite bare quartic coupling. For each value of m_0, y_0 the achieved heavy top quark mass $m_{t'}$ was evaluated to make sure that its values stayed within 3% with respect to 676 GeV. Finally the simulation was performed on various lattice volumes with identical parameters. Table 5.1 shows the chosen bare parameters and the obtained auto correlation time which was computed with the so called gamma method described in [57].

Figure 5.1 shows finite size effects of the scalar vev, the Higgs boson propagator and the fermion mass as well as the infinite volume extrapolations. The first row corresponds to vanishing bare quartic coupling and the second row shows the behaviour for infinite bare quartic coupling. The results for the observables are plotted against the inverse squared spatial lattice extent. The infinite volume extrapolation is performed using a linear fit involving lattice sizes of at least $L = 16$. The intercept of the linear fit at vanishing inverse squared lattice extent is then taken as the infinite volume result of the observable.

It is the aim to keep the quark masses fixed while the cut off is varied. In numerical simulations, where the result of observables involve statistical errors, it is a difficult task

Table 5.1: The table shows an overview of the chosen bare parameters. The fermion mass was evaluated at each value of the cut off in order to ensure, that the fermion mass is within 3% with respect to the average fermion mass of 676 GeV. *Stat.* is the number of configurations produced on a $16^3 \times 32$ lattice with the given parameter set. τ is the auto correlation time. The temporal extent aT was set to 32 in all cases.

κ	m_0^2/a^2	λ_0	y_0	$L_s\ [a \cdot L]$	Stat.	τ	Λ [GeV]
0.09442	2.59098	0.0	3.21224	12, 16, 18, 20, 24	20000	1.3	3498 ± 48
0.09463	2.56747	0.0	3.20867	12, 16, 18, 20, 24	15000	1.2	2929 ± 27
0.09485	2.54296	0.0	3.20495	12, 16, 18, 20, 24	15000	0.8	2548 ± 22
0.09545	2.47669	0.0	3.19486	12, 16, 18, 20, 24	20000	0.7	1883 ± 16
0.09560	2.46025	0.0	3.19235	12, 16, 18, 20, 24	20000	0.8	1786 ± 18
0.09605	2.41124	0.0	3.18486	12, 16, 18, 20, 24	20000	1.5	1511 ± 20
0.21300	$-\infty$	∞	3.37068	12, 16, 18, 20, 24	6000	7	3566 ± 48
0.21500	$-\infty$	∞	3.35497	12, 16, 18, 20, 24	6000	3	2701 ± 17
0.22200	$-\infty$	∞	3.18159	12, 16, 20, 24	15000	4	2563 ± 38
0.22320	$-\infty$	∞	3.17303	12, 16, 20, 24	15000	3	2299 ± 21
0.22560	$-\infty$	∞	3.15610	12, 16, 20, 24	15000	4	1932 ± 14
0.23040	$-\infty$	∞	3.12305	12, 16, 20, 24, 32	15000	3	1516 ± 7

to satisfy such a condition to arbitrary good precision. In this work the quark masses are about $m'_t = m'_b = 676 \pm 22$ GeV. Figure 5.2 shows the infinite volume result of the quark masses at various cut off values. The corresponding result for the standard model case is also presented.

Finally figure 5.3 shows the infinite volume result for the Higgs boson mass bounds at cut off values between 1500 GeV and 3500 GeV.

It is known from [48] that the upper Higgs boson mass bound follows the functional form

$$\frac{m_H^{up}}{a} = A_m \cdot \left\{ \log(\Lambda^2/\mu^2) + B_m \right\}^{-1/2}. \tag{5.6}$$

A_m, B_m are free fit parameters and μ is an arbitrary scale, which is set to $\mu = 1$ TeV.

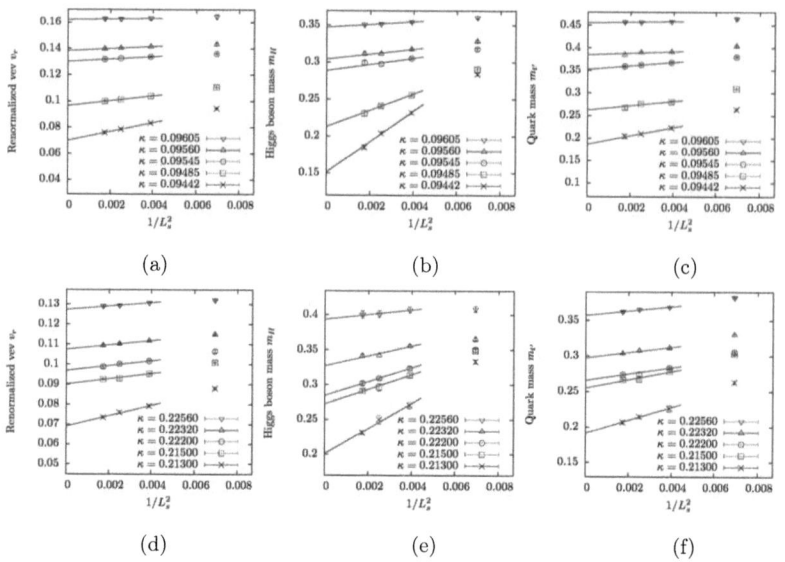

Figure 5.1: The figure shows the finite size effects and the infinite volume extrapolation of the renormalized vev, the mass from the pole of the Higgs boson propagator and the quark masses. The first row corresponds to vanishing bare quartic coupling and the second row is computed at infinite bare quartic coupling. The obtained results of the observables are plotted against the inverse squared spatial extent of the underlying lattice simulation. The red straight line is a linear fit of the data starting from a lattice extent of $L = 16$. The intercept of the linear fit with the vertical axis is taken as the infinite volume result.

Compared to the standard model case, the relative shift in the upper Higgs boson mass is less than 200 GeV and the cut off dependence is weaker. The bound is well compatible with the logarithmic decay given in (5.6). All parameter sets have been computed on lattices with spatial extent of $L_s \in 12, 16, 18, 20, 24$ in order to analyze the finite volume effects and to perform an infinite volume extrapolation (see table 5.1 for details on the simulation).

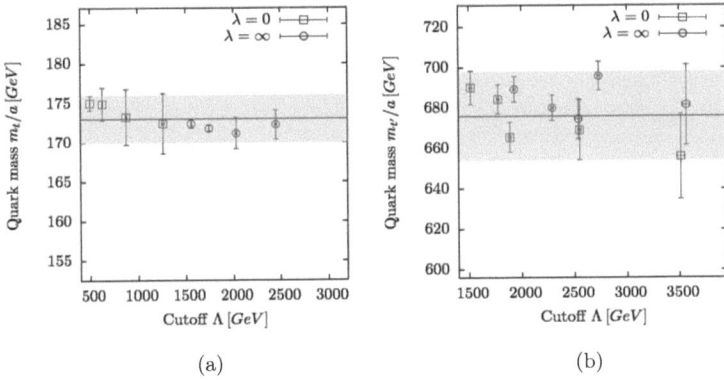

Figure 5.2: The figure shows the infinte volume extrapolation of the quark masses. The left plot (a) shows the result from the previous work [31, 30] in the standard model where the top quark mass was fixed at $m_t = m_b = 173 \pm 3$ GeV. The right plot (b) shows the case of a heavy quark doublet. The quark mass is $m'_t = m'_b = 676 \pm 22$ GeV.

5.5 Higgs boson mass bounds with varying top quark masses

The previous section aimed to attain top prime masses $(m_{t'})$ above the upper bound from partial wave unitarity considerations. In order to explore the heavy quark masses in the whole allowed region the cut off was kept constant at around 1.5 TeV and several top prime masses were simulated ranging from about 200 GeV up to 700 GeV. As observed in the previous chapter, the cut off effects as well as the finite size effects may play a significant role at a cut off of $\Lambda = 1500$ GeV and thus it will be mandatory to perform computations on large lattices such as 32^4.

As before, it is the aim to study the Higgs boson mass bounds. It turned out that the lower Higgs boson mass bound for large quark masses is numerically more demanding. The required computing time in order to reduce the relative statistical error of observables comparable to those for the upper bound is about a factor three larger. Due to restrictions in computing time and real time, only the upper Higgs boson mass bound with varying top prime masses was studied. Table 5.2 shows an overview of the chosen bare parameters,

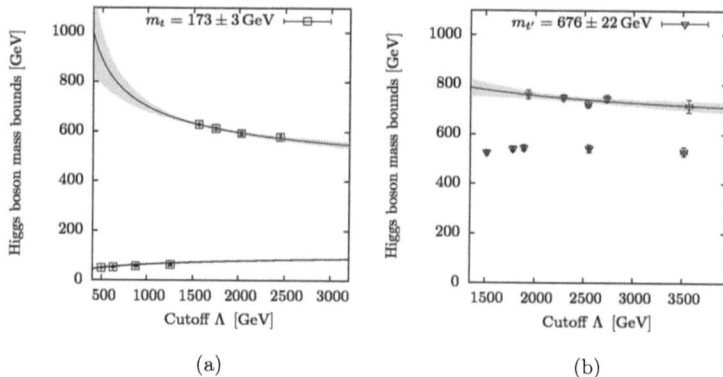

Figure 5.3: The figure shows the infinte volume extrapolation of the Higgs boson masses. The mass was extracted from the pole of the Higgs boson propagator. For a direct comparison, the results obtained in previous work [31, 30] is displayed in the left plot (a). The right plot (b) shows the case of a heavy quark doublet. The quark mass is $m'_t = m'_b = 676 \pm 22$ GeV.

the obtained cut off Λ and technical details about the simulations such as the number of produced configurations and the corresponding auto correlation time τ.

5.5.1 Numerical results

Table 5.2 shows the bare parameters of the model. At each value of the bare Yukawa coupling, the parameter κ or equivalently m_0^2 has to be tuned such that a cut off of about 1500 GeV is reached. Table 5.2 also shows the obtained cut off values on a $16^3 \times 32$ lattice. The final result for the cut off has to be taken after an extrapolation to infinite volume.

Figure 5.4 shows the finite size effects of the scalar vev, the Higgs boson mass and the quark masses. Though the infinite volume extrapolation of the scalar vev appears reliable, the finite size effects of the Higgs boson mass are strong and it seems to be necessary to perform the linear fits to infinite volume starting from lattice sizes of at least $L_s = 24$.

Finally figure 5.5 summarizes the final result on the Higgs boson mass bounds. In order to give a complete overview, the result from the previous section is repeated and displayed in the left image. The right image 5.5b shows the infinite volume result for the upper

Table 5.2: The table shows the chosen bare parameters, the obtained cut off Λ as well as the lattice sizes, the number of configurations and the auto correlation time τ.

m_0^2/a^2	λ_0	y_0	Lattice Extension $[a \cdot L]$	Stat. $16^3 \times 32$	τ	Λ [GeV]
$-\infty$	∞	3.12305	12, 16, 20, 24, 32	20000	3	1516 ± 7
$-\infty$	∞	2.04124	12, 16, 20, 24, 32	20000	2.8	1419 ± 4
$-\infty$	∞	1.30930	12, 16, 20, 24, 32	20000	2.6	1481 ± 3
$-\infty$	∞	0.96970	12, 16, 20, 24, 32	20000	1.6	1558 ± 4

Higgs boson mass. The mass is extracted from the pole of the propagator. In contrast to the previous case, the finite size effects of the Higgs boson mass is much stronger and therefore the extrapolation to infinite volume had to be performed with lattice volumes of at least $24^3 \times 32$.

5.6 Conclusion

The Higgs boson mass bounds have been studied in the presence of a heavy fourth generation of quarks. The model is restricted to the scalar sector and the heavy fermion doublet. The dominant contribution to the Higgs boson mass is expected to arise due to the large Yukawa coupling connected to heavy top prime quark. Hence, it seems acceptable to neglect all other contributions from particles which couple to the scalar sector within the standard model. Furthermore, the model is restricted to a degenerate fermion doublet which ensures that the fermion determinant is strictly real valued. The large Yukawa coupling necessitates a genuine non perturbative analysis which also respects the chiral nature of the electroweak model. Within this work, the basic physical observables are computed with the help of a polynomial hybrid Monte Carlo algorithm and the quark fields were incorporated with the Neuberger overlap operator which satisfies an exact chiral lattice symmetry.

An upper bound for a heavy top prime mass is suggested by perturbation theory on the basis of partial wave analysis and is about $m'_t = 550$ GeV. Hence, the cut off dependence

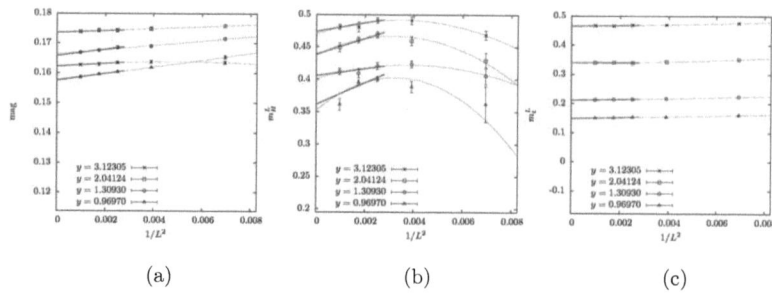

Figure 5.4: The plot shows the finite size effects and the infinite volume extrapolation of the scalar vev, the Higgs boson mass obtained from the pole of the propagator and the quark mass. The data is computed at infinite bare quartic coupling and thus corresponds to the upper Higgs boson mass. It was aimed to keep the cut off constant while the quark masses are varied between 200 GeV and 700 GeV.

of the Higgs boson mass bounds have been investigated in the presence of a top prime quark with a mass above the unitarity bound. The achieved top prime mass is around 670 GeV. In a second approach the upper Higgs boson mass bound has been computed at varying top prime quark masses and constant cut off $\Lambda = 1500$ GeV.

Figure 5.3b shows the upper and the lower Higgs boson mass bound in the presence of a heavy top prime quark with a mass of about 670 GeV. Figure 5.3a shows the analogous case for the standard model. The Higgs boson mass bounds are plotted against the cut off of the theory which is varied between 1500 GeV and 3000 GeV. All results were taken after an extrapolation to infinite volume. Figure 5.2 shows the finite size analysis of the scalar vev, the Higgs boson mass and the top quark mass. The first row corresponds to the lower Higgs boson mass bound where the bare quartic coupling is set to zero and the second row shows the case for the upper Higgs boson mass bound at infinite bare quartic coupling. The extrapolation to infinite volume are performed from lattice volumes of at least $16^3 \times 32$ and the infinite volume result is given by the intercept of the linear fit with the Y-Axis.

Similar to the case in the standard model, the upper bound is well compatible with the logarithmic decay given in equation (5.6). The presence of a heavy top quark has only mild effects on the upper Higgs boson mass bound. Compared to the standard model,

5.6 Conclusion

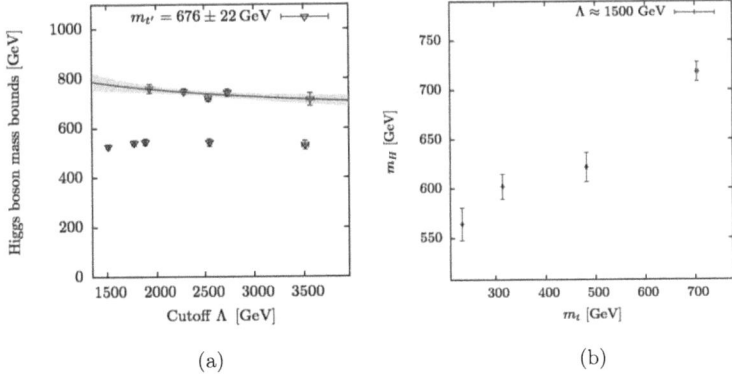

Figure 5.5: The figure summarizes the final result on the investigation of the Higgs boson mass bounds in the presence of a heavy fourth generation of quarks. The left plot shows the cut off dependent upper and lower Higgs boson mass bounds while the heavy quark mass was held fixed at around 676 ± 22 GeV. The right image (b) shows the upper Higgs boson mass bound at a constant cut off of about 1500 GeV while the heavy quark mass is varied between 200 GeV and 700 GeV.

the upper Higgs boson mass is shifted by about 25%. However, the lower Higgs boson mass is drastically raised by a factor of about five. The above results indicate that in the presence of a heavy top prime quark with a mass of about 670 GeV, Higgs boson masses below 500 GeV are excluded.

The strong dependence of the lower Higgs boson mass on the top prime mass motivates to investigate the dependence on top prime masses systematically. A first attempt is presented in figure 5.5 and shows the upper Higgs boson mass bound. The cut of is held fixed at around 1500 GeV while the top prime mass is varied between 200 GeV and 700 GeV. The figure also shows that the finite size effects are stronger for the chosen value of the cut off and the it is necessary to compute the Higgs boson mass on lattice volumes up to 32^4. The linear fit, from which the infinite volume result is extracted, involves lattice volumes of at least $20^3 \times 32$. As inferred from the investigation of the cut off dependence, the upper Higgs boson mass is influenced only mildly by a heavy top quark and the main contribution to the Higgs boson mass seem to be caused by the quartic scalar interaction.

It is certainly desirable to increase the number of configurations especially in the case for $y_0 = 0.96970$ where the finite size effects seem to be stronger than in the other cases. If one considers the result for the standard model case [31, 30] shown in figure 5.2, the Higgs boson mass is slightly above 600 GeV which deviates from the above mentioned point at $y_0 = 0.96970$ by about 2σ. Furthermore, the lower Higgs boson mass bound has not been investigated and is deferred to future work. As mentioned before, the lower Higgs boson mass bound is expected to have a strong dependence on the quark mass and is certainly of phenomenological interest. In this context it is appreciable to extend the analysis with higher dimensional terms in the action which may have a important influence on the lower mass bound. Another conceptually and physically interesting question is the highest possible top quark mass which is attainable within this model. The result should be contrasted to the upper bound of the top prime mass suggested by unitarity considerations.

6 Summary and conclusion

The Discovery of the Higgs boson would represent a further triumph for the standard model of elementary particle physics. With the upcoming experimental measurements at the LHC at Cern the chances for such a discovery are very high. However, the precise value of the Higgs boson mass cannot be provided by the standard model itself. Still, the theoretical tools within quantum field theory allow to set at least bounds on the Higgs boson mass and the corresponding widths.

Experimental measurements in the past decades could exclude a wide range of energies for the Higgs boson mass. The LEP experiments [9] state that the Higgs boson mass is above 114 GeV while preliminary data from the Tevatron experiments exclude a Higgs boson mass range between 158 GeV and 175 GeV [13]. The Higgs boson mass bounds were also addressed in perturbation theory. The perturbative mass bound of the upper Higgs boson mass relies on the triviality of the scalar sector, i.e. the renormalized quantities of the model necessarily vanish as the cut off approaches infinity. Hence, at each finite value of the cut off there is a maximal value for the renormalized quartic coupling which in turn implies an upper bound for the Higgs boson mass. Within perturbation theory the lower Higgs boson mass bound is derived from vacuum instability arguments. In both cases, it is not clear whether perturbation theory is applicable and therefore a genuine non perturbative analysis is necessary. The numerical method used in this work allows to compute the Higgs boson mass bounds from first principles. These bounds are also useful from a different point of view: once the Higgs boson mass is deduced from experimental measurements, the established numerical mass bounds can be used to infer the energy scale Λ where physics beyond the standard model has to play a significant role and should reveal signatures in high energy experiments. Furthermore, the numerical bounds on the Higgs boson mass may provide a stringent phenomenological criterium in order to distinguish the standard model Higgs boson from other not yet observed particles or bound states with the same quantum numbers.

The path integral formulation of quantum field theory allows to access physical observables numerically within the framework of lattice field theory. The latter approach offers a genuinely non perturbative too. The model under consideration in this thesis is the pure Higgs-Yukawa sector of the electroweak standard model which consists of a complex scalar Higgs doublet and a fermion doublet coupled to the scalar sector. The model closely resembles the weak interaction of the standard model. The chiral nature of the weak interaction is realized with the help of the Neuberger overlap lattice Dirac operator which enables to incorporate an exact chiral symmetry on the lattice.

It shall be remarked that the model does not contain gauge bosons and considers only a mass degenerate top quark doublet. Adding gauge fields in order to study a chiral gauge theory is still a challenge for the lattice formulation and goes far beyond the scope of this thesis. In particular, many important algorithmic improvements used in the simulation of the Higgs-Yukawa model cannot be transferred in the presence of gauge fields. Disregarding the gauge fields can be justified since the weak coupling constants are known to be small and thus the largest contribution to the Higgs boson mass is expected to arise from the Yukawa coupling between the Higgs boson and the top quark.

Within the past years large efforts have been undertaken to investigate the Higgs boson mass bounds in the framework of lattice field theory [31, 30, 20]. However, the decay width of the Higgs boson was not taken into account assuming that its effects are small. The main aim of this work is to treat the Higgs boson as a true resonance and to investigate the unstable nature of the Higgs boson from first principles. The resonance character of the Higgs boson manifests itself in the pole structure of the propagator. Chapter 2 and Chapter 3 therefore present a detailed discussion on the Higgs boson propagator and its pole structure. The calculation of the Higgs boson propagator within perturbation theory yields an analytic expression which is checked with numerical data to high precision. The full agreement of the perturbative expression and the numerical result, even at moderately large values of the Yukawa coupling, justifies to employ the perturbative functional form of the analytic propagator as a fit function for numerical data. The so obtained analytic fit function of the propagator then allows to compute the Higgs boson mass by locating the pole of the real part of the propagator in the complex plane. The influence of a non vanishing decay width which is connected to the imaginary part of the propagator is neglected. This procedure is compared with the Higgs boson mass obtained from the analysis of the scattering phases. The computation of the scattering phase within lattice field theory is based on the analysis of the volume dependence of energy eigenvalues [46]. Due to the finite volume of lattice simulations, the allowed lattice momenta are discrete. Hence, there are only a few two particle energy eigenvalues which are separated by distinct relative momenta and which lie in the elastic scattering region. In order to collect enough scattering phases near the resonance, it is therefore necessary to perform simulations on lattice volumes of at least $32^3 \times 40$. An extension of the finite size method has been proposed in [54] and is based on the analysis of the energy levels within a moving frame. This allows to compute scattering phases in the resonance region already on $24^3 \times 40$ lattices. For the present set of parameters, it turns out that the analysis of the moving frame energy eigenvalues are inevitable in order to extract the resonance parameters reliably. The centre of mass and the moving frame have been used for three different values of the quartic couplings. In all cases it was made sure that the Higgs boson can decay into two Goldstone particles and thus appears as a resonance. The simulations were performed on lattice volumes up to 40^4 and the analysis of the scattering phases in the rest frame as well as in the moving frame supplied enough data in order to perform a reliable Breit-Wigner fit near the resonance. The obtained results of the resonance mass and the width are presented in table 6.1. The results are also compared with those obtained from the analysis of the Higgs boson propagator and the energy eigenvalues obtained from a correlation matrix (GEVP) analysis [49, 11]. In all three cases the Higgs boson width is not larger than about 10% with respect to the resonance mass. Therefore, the corresponding total cross section exhibits a clear resonance peak even at the strongest value of the quartic coupling. Thus, the presented analysis gives quantitative results for the Higgs boson decay width and beyond that, it provides an a posteriori justification for the Higgs boson mass bounds in [31, 30].

Table 6.1 shows the first study of the dependence of Higgs boson resonance parameters on the quartic coupling based on first principles. The mass obtained from the propagator is in good agreement with the resonance mass obtained from the scattering phases. The

Table 6.1: The table summarizes the obtained final results on the resonance mass and the resonance width of the Higgs boson. $\hat{\lambda}$ denotes the bare quartic coupling. The first line is a preliminary result from Chapter 8 in [25]. Λ is the cut off of the theory. The following two columns display the resonance parameters computed from the scattering phases. Γ_H^p is the width obtained from perturbation theory where a non vanishing mass for the Goldstone bosons has been considered. Finally the mass extracted from the propagator as well as the mass eigenvalues computed with the help of the correlation matrix is shown. The latter results were obtained after an extrapolation to infinite volume as shown in figure 4.4 and figure 4.5.

$\hat{\lambda}$	Λ [GEV]	Res. mass M_H	Res. width Γ_H	Γ_H^p	Prop. mass M_H^p	GEVP
0.01	593(1)	0.428(3)	0.009(3)	0.0076(2)	0.433(3)	
0.01	883(1)	0.2811(6)	0.007(1)	0.0054(1)	0.278(2)	0.274(4)
1.0	1503(5)	0.374(4)	0.033(4)	0.036(8)	0.386(28)	0.372(4)
∞	1598(2)	0.411(3)	0.040(4)	0.052(2)	0.405(4)	0.403(7)

data in table 6.1 also shows that the analysis of the correlation matrix offers a reliable and simple method which is perfectly compatible with the previous methods explained above. Figure 6.1 summarizes the obtained total cross sections for the three different values of the bare quartic couplings. The displayed curves are the final fit results shown in figure 4.6.

Finally the simulation algorithm is used to explore the model at large Yukawa couplings in order to generate heavy fermions much above the standard model top quark mass. The existence of a heavy fourth generation of fermions is of phenomenological interest since it is not excluded by present high precision measurements. It provides various prospects to augment the model to enable a deeper understanding of flavour physics and mass hierarchies. It also serves as a source for CP violation. In addition it may give rise to a sufficiently strong first order phase transition such that the Sakharov conditions can be fulfilled in order to explain the observed Baryon asymmetry in the universe. It turns out that the upper and lower Higgs boson mass bounds indeed need to be revised in the presence of a heavy 700 GeV fermion doublet as studied here. The Higgs boson mass is extracted from the propagator as described in [31, 30]. The cut off is varied between 1500 GeV and 3500 GeV. Figure 6.2 summarizes the final results on the mass bounds and shows the corresponding results for the standard model top quark doublet. The results show that the upper bound is only slightly affected; the relative shift in the upper Higgs boson mass is about 200 GeV. The lower bound however, is drastically raised by a factor of about five to ten. The strong dependence of the lower bound on the heavy fermion mass motivates to investigate both bounds at varying top quark masses while the cut off is held fixed. A first computation has been performed at a cut off of 1500 GeV. The obtained results on the upper Higgs boson mass bound are presented in figure 5.5. The figure confirms the expected mild dependence of the upper Higgs boson mass on the Yukawa coupling. While the top quark mass is varied between 200 GeV and 700 GeV the effect on the upper Higgs boson mass is less than 25%. The established results severely restrict the mass range of a heavy fourth generation of fermions. For example, even if a Higgs boson of 400 GeV is found at the LHC, a heavy top quark with a mass of 700 GeV cannot be realized.

As mentioned before it will be interesting to explore especially the dependence of the lower Higgs boson mass bound on the heavy top prime mass. It turns out that the lower bound is numerically more demanding and therefore this is beyond the scope of this work. Nevertheless it is feasible and certainly of phenomenological interest. Another crucial question is the dependence of the lower Higgs boson mass bound on higher dimensional terms in the action. Furthermore, it is both of conceptual and of phenomenological importance to explore the highest possible fermion mass attainable within this model. The current bounds from partial wave unitarity conditions suggest a heavy top quark of at most 600 GeV. Top quark masses above this threshold value indicate non perturbative effects which are perfectly suitable to be explored with the numerical method on hand. It would be very interesting to check, whether the here considered Higgs-Yukawa model indeed exhibits a non perturbative regime which would immediately necessitate the framework of lattice field theory to reveal its properties.

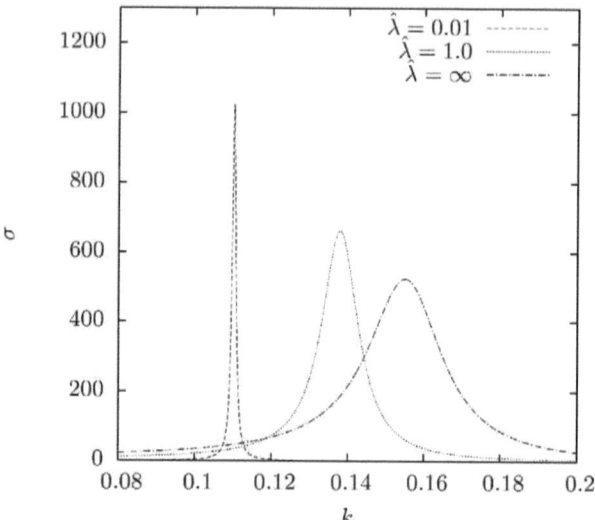

Figure 6.1: The figure shows the total cross section of the Higgs boson within the Higgs-Yukawa model. With regard to the standard model, this cross section is associated to the decay of the Higgs boson into the weak gauge bosons $W^\pm, Z$. The highest peak belongs to the smallest bare quartic coupling $\hat\lambda = 0.01$ and corresponds to a Higgs boson resonance mass of $M_H = 248 \pm 1$ GeV and the resonance width is $\Gamma_H = 6.2 \pm 0.9$ GeV. The next peak is obtained at $\hat\lambda = 1.0$ and corresponds to $M_H = 562 \pm 2$ GeV and $\Gamma_H = 50 \pm 6$ GeV. The last peak is associated to infinite bare quartic coupling and corresponds to $M_H = 618 \pm 5$ GeV and $\Gamma_H = 60 \pm 6$ GeV. The cut off value for the three different couplings is displayed in table 6.1. In all cases, the resonance width is less than 10% with respect to the resonance mass and thus the corresponding total cross section exhibits a clear resonance peak.

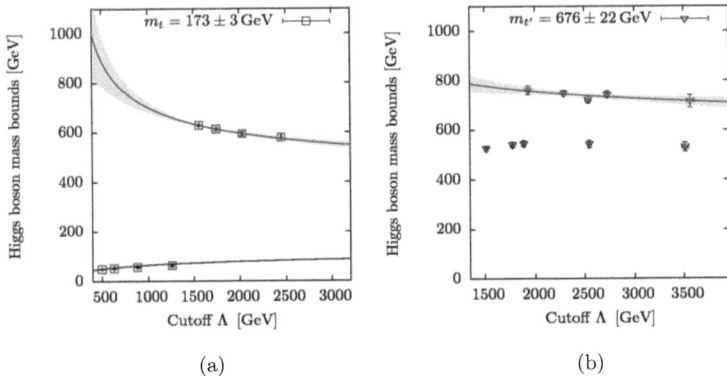

Figure 6.2: The figure shows the infinite volume extrapolation of the Higgs boson masses. The mass is extracted from the pole of the real part of the Higgs boson propagator. For a direct comparison, the results for the standard model top quark doublet is displayed in the left plot (a). The right plot (b) shows the case of a heavy quark doublet. The quark mass is about 700 GeV

Appendix A: Lattice Parametrization

The action was already given in (2.1). The scalar part of the action can be defined on a finite and discrete lattice by substituting the derivatives with finite differences and the integrals by a sum over the space time lattice. The fermionic part is more complicated and is discussed in Chapter 2. The calculations presented below show that the announced relation for the lattice parametrization of the Higgs-Yukawa model indeed resembles the continuum notation. The well known Lagrangian of the Higgs-Yukawa model is given in (2.1) and is repeated here

$$L_E^{HY} = \frac{1}{2}(\partial_\mu \varphi)^\dagger \cdot (\partial^\mu \varphi) + \frac{1}{2}m^2 \varphi^\dagger \cdot \varphi + \lambda \left(\varphi^\dagger \cdot \varphi\right)^2$$

$$+ \bar{t}\,\slashed{D}\,t + \bar{b}\,\slashed{D}\,b + y_b \begin{pmatrix}\bar{t}\\\bar{b}\end{pmatrix}_L^T \cdot \varphi\, b_R + y_t \begin{pmatrix}\bar{t}\\\bar{b}\end{pmatrix}_L^T \cdot \tilde{\varphi}\, t_R \quad + h.c..$$

The subscript E denotes the Euclidean version of the standard model electroweak sector. The action is then given by

$$S = \int d^4x\, L_E^{HY}\left(\partial_\mu \varphi(x), \varphi(x), \partial_\mu \psi(x), \psi(x)\right), \quad \psi(x) = \begin{pmatrix} t(x) \\ b(x) \end{pmatrix}$$

In order to clarify the lattice parametrization, it is not necessary to consider the fermion sector. Hence, the below calculations shall focus only on the scalar part. Furthermore, the calculations are straight forward in the case an N component scalar vector and therefore the calculations are restricted to the one component scalar theory.

The derivatives are substituted by finite differences $\partial_\mu \to \Delta_\mu^f$ where the finite difference operator is defined by

$$\Delta_\mu^f f(x) := \frac{1}{a}\{f(x + a\hat{\mu}) - f(x)\},$$

$$\Delta_\mu^b f(x) := \frac{1}{a}\{f(x) - f(x - a\hat{\mu})\}.$$

In the following, the scalar fields restricted to the lattice points shall be denoted by the same variable φ.

$$\Rightarrow S_\varphi = a^4 \sum_{x,\mu} \frac{1}{2} \left\{ \frac{1}{a^2} \Big(\varphi(x+a\hat{\mu}) - \varphi(x)\Big)\Big(\varphi(x+a\hat{\mu}) - \varphi(x)\Big) \right\} +$$
$$a^4 \sum_x \frac{1}{2} m_0^2 \varphi^2(x) + \lambda a^4 \sum_x \varphi^4(x)$$
$$= a^4 \sum_{x,\mu} \frac{1}{2a^2} \left\{ \varphi^2(x+a\mu) + \varphi^2(x) - 2\varphi(x+a\mu)\varphi(x) \right\} +$$
$$a^4 \sum_x \frac{1}{2} m_0^2 \varphi^2(x) + \lambda a^4 \sum_x \varphi^4(x)$$
$$= -a^4 \sum_{x,\mu} \frac{1}{a^2} \varphi(x+a\mu)\varphi(x) + 4 a^4 \sum_x \frac{1}{a^2} \varphi^2(x)$$
$$+ a^4 \sum_x \frac{1}{2} m_0^2 \varphi^2(x) + \lambda a^4 \sum_x \varphi^4(x)$$

For numerical simulation it is common to rescale the field with a parameter κ such that

$$\phi = \sqrt{2\kappa}\, \varphi,\, a \equiv 1$$

$$= -2\kappa \sum_{x,\mu} \phi(x+a\mu)\phi(x) + \sum_x \left\{ 8\phi^2(x) + m_0^2\, \kappa\, \phi^2(x) + 4\lambda \kappa^2 \phi^4(x) \right\}$$
$$= -2\kappa \sum_{x\mu} \phi(x+a\mu)\phi(x) + \sum_x \left\{ (8+m_0^2)\, \kappa\, \phi^2(x) + 4\lambda\kappa^2 \phi^4(x) \right\}$$
$$\Rightarrow \hat{\lambda} = 4\lambda\kappa^2 \quad \Rightarrow \lambda = \frac{\hat{\lambda}}{4\kappa^2}$$
$$= \sum_x \left\{ -2\kappa \sum_\mu \phi(x+a\mu)\phi(x) + \hat{\lambda}\left(\frac{(8+m_0^2)\,\kappa}{\hat{\lambda}} \phi^2(x) + \phi^4(x) \right) \right\}$$

κ is a free parameter and can be chosen such that the last line in the above equation can be simplified

$$\frac{(8+m_0^2)\kappa}{\hat{\lambda}} = -2 \quad \Rightarrow m_0^2 = \frac{-2\hat{\lambda}}{\kappa} - 8$$

$$\Rightarrow S = \sum_x \left\{ -2\kappa \sum_\mu \phi(x+a\mu)\phi(x) + \hat{\lambda}\left(-2\,\phi^2(x) + \phi^4(x) + 1 - 1\right) \right\}$$
$$= \sum_x \left\{ -2\kappa \sum_\mu \phi(x+a\mu)\phi(x) + \hat{\lambda}\left(\left(\phi^2(x) - 1\right)^2 - 1 \right) \right\}$$
$$= \sum_x \left\{ -2\kappa \sum_\mu \phi(x+a\mu)\phi(x) + \hat{\lambda}\left(\phi^2(x) - 1\right)^2 - \hat{\lambda} \right\}$$

However, it is common to choose κ such that

$$\frac{(8+m_0^2)\kappa - 1}{\hat{\lambda}} = -2 \quad \Rightarrow m_0^2 = \frac{1 - 2\hat{\lambda} - 8\kappa}{\kappa}$$

In that that case the action is

$$S = \sum_x \left\{ -2\kappa \sum_\mu \phi(x+a\mu)\phi(x) + \phi^2(x) + \hat{\lambda}[\phi^2(x) - 1]^2 - \hat{\lambda} \right\}.$$

Appendix B: Perturbative calculations

This appendix collects some perturbative calculations in detail. The below calculations will need the use of Feynman parameterization and some solutions of trigonometric and logarithmic integrals. To keep the argumentation free of technical details, some mathematical identities are elaborated at the beginning.

The Feynman parametrization of loop integrals lead to integrals of the following types:

1. $\int dx \, \frac{1}{-x^2+x+c}$

2. $\int dx \, \ln(-x^2 + ax + c)$

3. $\int dx \, \frac{x^2}{-x^2+x+c}$

These are elementary integrals and can be solved easily. The results and some steps are summarized below.

- $C_1(x) = \int dx \, \frac{1}{-x^2+x+c}$

    First we rewrite the denominator as
    $$\frac{1}{-x^2+x+c} = \frac{1}{\eta^2 - y^2}$$
    $$\eta^2 = \frac{1}{4} + c, \, y = x - \frac{1}{2}.$$

    Decomposing the fraction into parts leads to
    $$\frac{1}{\eta^2 - y^2} = \frac{1}{2\eta}\left(\frac{1}{y+\eta} + \frac{1}{\eta-y}\right)$$
    $$= \frac{1}{2\eta}\frac{d}{dy}\left(\ln(\eta+y) - \ln(\eta-y)\right)$$
    $$= \frac{d}{dy}\frac{1}{2\eta}\ln\left(\frac{\eta+y}{\eta-y}\right).$$

So the indefinite integral can be written as

$$C_1(x) = \frac{1}{\sqrt{1+4c}} \log\left(\frac{\sqrt{1+4c} + (2x-1)}{\sqrt{1+4c} - (2x-1)}\right)$$

Special care is now needed in order to evaluate the indefinite integral in the corresponding limits, as the logarithm as well as the the square root may need to be analytically continued into the complex plane. The $i\epsilon$ prescription of the propagator is then needed to decide on which Riemann surface the log has to be evaluated. Further, the well known rules about the real logarithms are not valid for complex arguments. The upper limit $x = 1$ induces a negative sign in the denominator of the argument of the logarithm. On the other hand the numerator is negative for the lower limit $x = 0$.

The definite integral in the limits $x = 0$ to $x = 1$ is then

$$\boxed{\Delta C_1 = \frac{1}{\sqrt{\frac{1}{4}+c}} \ln\left(\frac{\frac{1}{2} + \sqrt{\frac{1}{4}+c}}{-\frac{1}{2} + \sqrt{\frac{1}{4}+c}}\right).}$$

- $C_2(x) = \int dx \, \ln(-x^2 + ax + c)$

$$j_2(x) := \ln(-x^2 + x + c)$$
$$= \ln\left(-y^2 + \eta^2\right); \quad y := x - \frac{a}{2}, \quad \eta^2 = \frac{a^2}{4} + c.$$

$$j_2(y) = \frac{d}{dy}\left(y \ln(\eta^2 - y^2)\right) + \frac{2y^2}{\eta^2 - y^2}$$
$$= \frac{d}{dy}\left(y \ln(\eta^2 - y^2)\right) - 2\left(\frac{-y^2}{\eta^2 - y^2} + \frac{\eta^2}{\eta^2 - y^2} - \frac{\eta^2}{\eta^2 - y^2}\right)$$
$$= \frac{d}{dy}\left(y \ln(\eta^2 - y^2)\right) - 2\left(1 - \frac{\eta^2}{\eta^2 - y^2}\right)$$
$$= \frac{d}{dy}\left(y \ln(\eta^2 - y^2) - 2y\right) + 2\eta^2 \left(\frac{1}{\eta^2 - y^2}\right)$$
$$= \frac{d}{dy}\left(y \ln(\eta^2 - y^2) - 2y\right) + 2\eta^2 \frac{1}{2\eta} \frac{d}{dy} \ln\left(\frac{\eta + y}{\eta - y}\right)$$
$$= \frac{d}{dy}\left\{y \ln(\eta^2 - y^2) - 2y + \eta \ln\left(\frac{\eta + y}{\eta - y}\right)\right\}$$

Inserting the definition of η and y we get in the boundaries $x \in [0, 1]$

$$\Delta C_2 = \frac{1}{2}\Big\{(a\log(c) - (a-2)\log(a+c-1)$$
$$+ \sqrt{a^2+4c}\,\log\left(\frac{a - \sqrt{a^2+4c} - 2}{a + \sqrt{a^2+4c} - 2}\right)$$
$$- \sqrt{a^2+4c}\,\log\left(\frac{a - \sqrt{a^2+4c}}{a + \sqrt{a^2+4c}}\right) - 4\Big\}$$

The special case $a = 1$ will be useful which shall be denoted by

$$\Delta C_2^1 := \Delta C_2(a=1) = -2 + \log(c) + \sqrt{4c+1}\,\log\left(\frac{\sqrt{4c+1}+1}{\sqrt{4c+1}-1}\right)$$

- $C_3(x) = \int dx\, (-x^2 + x + c)\log(-x^2 + x + c)$

As before we can rewrite the integral with η and y and we get

$$j_3(y) = \eta^2 \ln\left(\eta^2 - y^2\right) - y^2 \ln\left(\eta^2 - y^2\right)$$
$$= \eta^2\, j_2(y) - y^2 \ln\left(\eta^2 - y^2\right)$$

$$-y^2 \ln\left(\eta^2 - y^2\right) = \frac{d}{dy}\left\{-\frac{1}{3}y^3 \ln\left(\eta^2 - y^2\right)\right\} - \frac{2y^4}{3(\eta^2 - y^2)}$$
$$= dy\Big\{-\frac{1}{3}y^3 \ln\left(\eta^2 - y^2\right)$$
$$- \frac{1}{9}\left(3\eta^3 \ln\left(\frac{y+\eta}{y-\eta}\right) - 2\left(y^3 + 3\eta^2 y\right)\right)\Big\}$$

$$\Rightarrow j_3(y) = \frac{1}{9}\Big(6\log\left(\frac{y+\eta}{y-\eta}\right)\eta^3 + 2\left(y^3 - 6y\eta^2\right)$$
$$- 3\left(y^3 - 3y\eta^2\right)\log\left(\eta^2 - y^2\right)\Big).$$

In the given limits, the above integral is

$$\Delta C_3 = \frac{1}{3}\Big((6c+1)\log(c)$$
$$- \frac{1}{2}(4c+1)^{\frac{3}{2}}\log\left(-\frac{\sqrt{4c+1}-1}{\sqrt{4c+1}+1}\right) - \left(4c+\frac{5}{6}\right)\Big).$$

The three types of loop integrals I_1, I_2, J will be discussed. As the loop contributions involves products of propagators, the identity

$$\frac{1}{A^a}\frac{1}{B^b} = \frac{\Gamma(a+b)}{\Gamma(a)\Gamma(b)} \int_0^1 dx\, \frac{x^{a-1}\bar{x}^{b-1}}{(xA+\bar{x}B)^{a+b}}, \quad \bar{x} = 1-x$$

will be used. Γ is the Gamma function. Scalar one loop integrals have the form

$$\tilde{A}(a,\xi) := \int \frac{\mathrm{d}^4 k}{(2\pi)^4} \frac{1}{(k^2 - \xi + i\epsilon)^a}$$

After performing a Wick-Rotation:

$$\tilde{A}(a,\xi) = (-1)^a i \int \frac{\mathrm{d}^4 k}{(2\pi)^4} \frac{1}{(k^2 + \xi)^a}$$

Where k is now a Euclidean four vector. The following calculations are entirely performed in Euclidean space time. It is therefore useful to define

$$A(a,\xi) = \int \frac{\mathrm{d}^4 k}{(2\pi)^4} \frac{1}{(k^2 + \xi)^a}.$$

The integral only depends on the norm of the 4 dimensional vector, expressed in spherical coordinates it reads

$$\int \mathrm{d}^4 k \longrightarrow \pi^2 \int_0^\infty \mathrm{d}k^2 \, k^2.$$

And so

$$A(a,\xi) = \frac{1}{(4\pi)^2} \int_0^\infty \mathrm{d}k^2 \, \frac{k^2}{(k^2 + \xi)^a}$$

The special cases $a = 1, 2$ will be needed later.

- $\boxed{a = 1}$

  Define:

  $$I(\xi, L) := \int_0^L \mathrm{d}z \, \frac{z}{z + \xi}$$

  The indefinite integral gives

  $$\int \mathrm{d}z \, \frac{z}{z + \xi} = z - \xi \ln(z + \xi)$$

  $$\frac{z}{z+\xi} = \frac{z}{z+\xi} + \frac{\xi}{z+\xi} - \frac{\xi}{z+\xi} \tag{B1}$$

  $$= 1 - \frac{\xi}{z+\xi} \tag{B2}$$

  $$= \frac{\mathrm{d}}{\mathrm{d}z}(z - \xi \ln(z + \xi)) \tag{B3}$$

  $$\Rightarrow I(\xi, L) = (z - \xi \ln(z + \xi))\Big|_{z=0}^{z=L} \tag{B4}$$

  $$= L - \xi \ln(L + \xi) + \xi \ln(\xi) \tag{B5}$$

  $$= L + \xi \ln\left(\frac{\xi}{L + \xi}\right) \tag{B6}$$

L is the regulator of the integral and will be referred to as the cut off Λ:

$$\boxed{A(1,\xi,\Lambda^2) = \frac{1}{(4\pi)^2}\left\{\Lambda^2 - \xi\ln\left(\frac{\Lambda^2+\xi}{\xi}\right)\right\}}$$

- $\boxed{a=2}$

$$A(2,\xi,\Lambda^2) = \frac{1}{(4\pi)^2}\int_0^{\Lambda^2} dk^2\, \frac{k^2}{(k^2+\xi)^2}$$

The integrand can be rewritten $g(z) = \frac{z}{(z+\xi)^2}$

$$g(z) = \frac{z}{(z+\xi)^2} = -z\frac{d}{dz}\frac{1}{z+\xi} \tag{B7}$$

$$= -\frac{d}{dz}\left(\frac{z}{z+\xi}\right) + \frac{1}{z+\xi} \tag{B8}$$

$$= \frac{d}{dz}\left(-\frac{z}{z+\xi} + \ln(z+\xi)\right) \tag{B9}$$

$$\tag{B10}$$

The indefinite integral is then given by

$$G(z) := -\frac{z}{z+\xi} + \ln(z+\xi) \tag{B11}$$

$$\Rightarrow G(L) - G(0) = \ln(L+\xi) - \frac{L}{L+\xi} - \ln(\xi) \tag{B12}$$

$$= \ln\left(\frac{L+\xi}{\xi}\right) - \frac{L}{L+\xi} \tag{B13}$$

$$\Rightarrow \boxed{A(2,\xi,\Lambda^2) = \frac{1}{(4\pi)^2}\left\{\ln\left(\frac{\Lambda^2+\xi}{\xi}\right) - \frac{\Lambda^2}{\Lambda^2+\xi}\right\}}$$

It is common to give the result containing only the leading terms in $\frac{\xi}{\Lambda^2}$. Therefore the argument of the logarithms have to be expanded

$$\frac{\xi}{\Lambda^2+\xi} = \frac{\frac{\xi}{\Lambda^2}}{1+\frac{\xi}{\Lambda^2}} \tag{B14}$$

$$\frac{x}{1+x} := \approx 0 + \left(\frac{1}{1+x} - \frac{x}{(1+x)^2}\right)\bigg|_{x=0} x = x + \mathcal{O}(x^2) \tag{B15}$$

$$\Rightarrow \frac{\xi}{\Lambda^2+\xi} = \frac{\xi}{\Lambda^2} + \mathcal{O}\left(\left(\frac{\xi}{\Lambda^2}\right)^2\right) \tag{B16}$$

Analogously:
$$\frac{\Lambda^2}{\Lambda^2+\xi} \approx 1 - \frac{\xi}{\Lambda^2} \tag{B17}$$

the fermion loop involves an integral of the follwing type
$$B(a,\xi,\Lambda^2) := \int \frac{d^4q}{(2\pi)^4} \frac{q^2}{(q^2+\xi)^a}, \qquad a=2.$$

The procedure analogous to the calculations given for the evaluation of the integrals $A(n,\xi,\lambda^2)$.

The above relations will be useful to calculate the one loop scalar and fermion contributions. Finally the main results shall be summarized:

$$A(1,\xi,\Lambda^2) = \int \frac{d^4k}{(2\pi)^4} \frac{1}{(k^2+\xi)}$$
$$= \frac{1}{(4\pi)^2}\left\{\Lambda^2 - \xi \ln\left(\frac{\Lambda^2+\xi}{\xi}\right)\right\}$$
$$\approx \frac{1}{(4\pi)^2}\left\{\Lambda^2 - \xi \ln\left(\frac{\Lambda^2}{\xi}\right)\right\}$$

$$A(2,\xi,\Lambda^2) = \int \frac{d^4k}{(2\pi)^4} \frac{1}{(k^2+\xi)^2}$$
$$= \frac{1}{(4\pi)^2}\left\{\ln\left(\frac{\Lambda^2+\xi}{\xi}\right) - \frac{\Lambda^2}{\Lambda^2+\xi}\right\}$$
$$\approx \frac{1}{(4\pi)^2}\left\{\ln\left(\frac{\Lambda^2}{\xi}\right) - 1\right\}$$

$$B(2,\xi,\Lambda^2) = \int \frac{d^4q}{(2\pi)^4} \frac{q^2}{(q^2+\xi)^2}$$
$$= \frac{1}{(4\pi)^2}\left\{\frac{\Lambda^2(\Lambda^2+2\xi)}{\Lambda^2+\xi} + 2\xi \log\left(\frac{\xi}{\Lambda^2+\xi}\right)\right\}$$
$$\approx \frac{1}{(4\pi)^2}\left\{\Lambda^2 + 2\xi \log\left(\frac{\xi}{\Lambda^2}\right)\right\}$$

The scalar loop integral with a single scalar mass is denoted by I_1 and is given by Σ_2 is given by:

$$I_1\left(p^2,m^2,\Lambda\right) := \int \frac{d^4q}{(2\pi)^4} \frac{1}{q^2+m^2} \frac{1}{(p+q)^2+m^2}, \qquad q^0 < \Lambda$$

Introducing Feynman parameters, the integral is rewritten as

$$I\left(p^2, m^2, \Lambda\right) = \int_0^1 dx \int \frac{d^4q}{(2\pi)^4} \frac{1}{(x(q^2 + m^2) + \bar{x}((p+q)^2 + m^2))^2}$$

$$= \int_0^1 dx \int \frac{d^4q}{(2\pi)^4} \frac{1}{(q^2 + p^2\bar{x} + 2pq\bar{x} + m^2)^2}$$

$$= \int_0^1 dx \int \frac{d^4q}{(2\pi)^4} \frac{1}{((q+p\bar{x})^2 - p^2\bar{x}^2 + p^2\bar{x} + m^2)^2}$$

$$= \int_0^1 dx \int \frac{d^4q}{(2\pi)^4} \frac{1}{((q+p\bar{x})^2 + m^2 + p^2 x\bar{x})^2}$$

Performing a shift in the integration variable q

$$q \longrightarrow q + p\bar{x}$$

yields

$$\Rightarrow I_1\left(p^2, m^2, \Lambda\right) = \int_0^1 dx \int \frac{d^4q}{(2\pi)^4} \frac{1}{(q^2 + M^2(p^2, m^2, x))^2}$$

$$M^2(p^2, m^2, x) := m^2 + p^2 x\bar{x}$$

$$\Rightarrow I_1(p^2; m^2, \Lambda^2) = \int_0^1 dx \, A(2, M^2, \Lambda^2)$$

$$= \frac{1}{(4\pi)^2} \int_0^1 dx \left\{ \ln\left(\frac{\Lambda^2 + M^2}{M^2}\right) - \frac{\Lambda^2}{\Lambda^2 + M^2} \right\}$$

$$\approx \frac{1}{(4\pi)^2} \int_0^1 dx \left\{ \ln\left(\frac{\Lambda^2}{M^2}\right) - 1 \right\}$$

It is sufficient to solve the integral over x by taking only the leading term in $\frac{M^2}{\Lambda^2}$

$$I_1(p^2, m^2, \Lambda^2) := \frac{1}{(4\pi)^2} \int_0^1 dx \left\{ \ln\left(\frac{\Lambda^2}{M^2(p^2, m^2, x)}\right) - 1 \right\}$$

$$= \frac{1}{(4\pi)^2} \int_0^1 dx \left\{ \ln\left(\frac{\Lambda^2}{m^2 + p^2 x \bar{x}}\right) - 1 \right\}$$

$$= \frac{1}{(4\pi)^2} \int_0^1 dx \left\{ \ln\left(\frac{\Lambda^2}{m^2 + p^2 x - p^2 x^2}\right) - 1 \right\}$$

$$= \frac{1}{(4\pi)^2} \int_0^1 dx \left\{ \ln\left(\frac{\Lambda^2}{p^2(\frac{m^2}{p^2} + x - x^2)}\right) - 1 \right\}$$

$$= \frac{1}{(4\pi)^2} \int_0^1 dx \left\{ \ln\left(\frac{\Lambda^2}{p^2}\right) - 1 + \ln\left(\frac{1}{\frac{m^2}{p^2} + x - x^2}\right) \right\}$$

$$= \frac{1}{(4\pi)^2} \left\{ \ln\left(\frac{\Lambda^2}{p^2}\right) - 1 - \int_0^1 dx \ln\left(-x^2 + x + \frac{m^2}{p^2}\right) \right\}$$

$$= \frac{1}{(4\pi)^2} \left\{ \ln\left(\frac{\Lambda^2}{p^2}\right) - 1 - \Delta I_2\left(a = 1, c = \frac{m^2}{p^2}\right) \right\}$$

$$= \frac{1}{(4\pi)^2} \left\{ \ln\left(\frac{\Lambda^2}{p^2}\right) - 1 \right.$$

$$\left. - \left(-2 + \ln(\frac{m^2}{p^2}) + \sqrt{1 + \frac{4m^2}{p^2}} \ln\left(\frac{\sqrt{1 + \frac{4m^2 - i\epsilon}{p^2}} + 1}{\sqrt{1 + \frac{4m^2 - i\epsilon}{p^2}} - 1}\right)\right) \right\}$$

$$I_1(p^2, m^2, \Lambda^2) = \frac{1}{(4\pi)^2} \left\{ 1 + \ln\left(\frac{\Lambda^2}{m^2}\right) \right.$$

$$\left. - \sqrt{1 + \frac{4m^2}{p^2}} \ln\left(\frac{1 + \sqrt{1 + \frac{4m^2}{p^2}} - i\epsilon \, \text{Sgn}(p^2)}{-1 + \sqrt{1 + \frac{4m^2}{p^2}} - i\epsilon \, \text{Sgn}(p^2)}\right) \right\}$$

The mixed scalar loop integral which arises in the self energy contribution of the Goldstone boson is defined by

$$I_2(p^2, m_\varphi^2, m_G^2, \Lambda) = \int \frac{d^4q}{(2\pi)^4} \frac{1}{q^2 + m_G^2} \frac{1}{(p+q)^2 + m_\phi^2}, \qquad q^0 < \Lambda.$$

Again introducing Feynman parameters, the denominator can be written in the form suitable to use to the integral expressions given above

$$I_2(p^2, m_\varphi^2, m_G^2, \Lambda) = \int_0^1 dx \int \frac{d^4q}{(2\pi)^4} \frac{1}{\left\{x(q^2 + m_G^2) + \bar{x}\left((p+q)^2 + m_\phi^2\right)\right\}^2}.$$

After expanding the denominator, define $Q := q + p\bar{x}$ and the integral can be expressed in the shifted variable Q. The boundary terms that arise due to the finite cut off $q^0 < \Lambda$

are neglected

$$I_2(p^2, m_\varphi^2, m_G^2, \Lambda) = \int_0^1 dx \int \frac{d^4Q}{(2\pi)^4} \frac{1}{\left\{Q^2 + m_\phi^2 - p^2 x^2 + \left(m_G^2 - m_\phi^2 + p^2\right) x\right\}^2}.$$

With

$$\xi = m_\phi^2 + (m_G^2 - m_\phi^2 + p^2)x - p^2 x^2$$

I_2 takes the form

$$\begin{aligned}I_2(p^2, m_\varphi^2, m_G^2, \Lambda) &= \int_0^1 dx\, A(2, \xi, \Lambda^2) \\ &= \frac{1}{16\pi^2} \int_0^1 dx \left\{ \ln\left(\frac{\Lambda^2}{-p^2 x^2 + \left(p^2 + m_G^2 - m_\phi^2\right)x + m_\phi^2}\right) - 1 \right\} \\ &= \frac{1}{16\pi^2} \int_0^1 dx \left\{ \ln\left(\frac{\Lambda^2}{p^2} \frac{1}{-x^2 + \left(\frac{m_G^2}{p^2} - \frac{m_\phi^2}{p^2} + 1\right) x + \frac{m_\phi^2}{p^2}}\right) - 1 \right\} \\ &= \frac{1}{16\pi^2} \left\{ \ln\left(\frac{\Lambda^2}{p^2}\right) - 1 \right. \\ &\qquad \left. - \int_0^1 dx \ln\left(-x^2 + \left(\frac{m_G^2}{p^2} - \frac{m_\phi^2}{p^2} + 1\right) x + \frac{m_\phi^2}{p^2}\right) \right\}.\end{aligned}$$

Setting

$$a = \frac{m_G^2}{p^2} - \frac{m_\phi^2}{p^2} + 1,$$
$$c = \frac{m_\phi^2}{p^2}$$

yields

$$I_2(p^2, m_\varphi^2, m_G^2, \Lambda) = \frac{1}{16\pi^2} \left\{ \ln\left(\frac{\Lambda^2}{p^2}\right) - 1 - \Delta C_2 \right\}.$$

After substituting the variables a, c finally the result is

$$\begin{aligned}I_2(p^2, m_\varphi^2, m_G^2, \Lambda) &= \frac{1}{32\pi^2} \left\{ 2 + \log\left(\frac{\Lambda^4}{m_G^2 m_\varphi^2}\right) + \frac{(m_G^2 - m_\varphi^2)}{p^2} \log\left(\frac{m_G^2}{m_\varphi^2}\right) \right. \\ &\qquad \left. + \log\left(\frac{p^2 + m_G^2 + m_\varphi^2 - p^2\, \kappa(p^2) + i\epsilon\, \text{Sgn}\,(p^2)}{p^2 + m_G^2 + m_\varphi^2 + p^2\, \kappa(p^2) - i\epsilon\, \text{Sgn}\,(p^2)}\right) \kappa(p^2) \right\}. \\ \kappa^2(p^2) &:= \frac{4\, p^2 m_\varphi^2 + \left(p^2 + m_G^2 - m_\varphi^2\right)^2}{p^4}\end{aligned}$$

The fermion loop is given by

$$\begin{aligned}
J(p^2, m_f, \Lambda) &= \int \frac{d^4q}{(2\pi)^4} Tr\left[\frac{1}{(\slashed{p}+\slashed{q}+m)}\frac{1}{(\slashed{q}+m)}\right], \quad q^0 < \Lambda \\
&= \int \frac{d^4q}{(2\pi)^4} \frac{Tr\left[(\slashed{p}+\slashed{q}-m)(\slashed{q}-m)\right]}{((p+q)^2 - m^2)(q^2 - m^2)} \\
&= 4\int_0^1 dx \int \frac{d^4q}{(2\pi)^4} \frac{q^2 + m^2 + pq}{((q+px)^2 + p^2 x\bar{x} - m^2)^2}, \quad q \to q + px \\
&= 4\int_0^1 dx \int \frac{d^4q}{(2\pi)^4} \frac{q^2 - p^2 x\bar{x} + m^2}{(q^2 + p^2 x\bar{x} - m^2)^2}, \quad M^2(p^2, x) := p^2 x\bar{x} - m^2 \\
&= 4\int_0^1 dx \int \frac{d^4q}{(2\pi)^4} \frac{q^2 - M^2(p^2, x)}{(q^2 + M^2(p^2, x))^2} \\
&= 4\int_0^1 dx \left\{ \int \frac{d^4q}{(2\pi)^4} \frac{q^2}{(q^2 + M^2(p^2, x))^2} \right. \\
&\quad \left. - M^2(p^2, x) \int \frac{d^4q}{(2\pi)^4} \frac{1}{(q^2 + M^2(p^2, x))^2} \right\} \\
&= 4\int_0^1 dx \left\{ B(2, M^2, \Lambda^2) - M^2 A(2, M^2, \Lambda^2) \right\} \\
&= \frac{1}{4\pi^2} \int_0^1 dx \left(\Lambda^2 + M^2 + 3 M^2 \log\left(\frac{M^2}{\Lambda^2}\right) \right) \\
&= \frac{1}{4\pi^2} \left\{ \Lambda^2 + \frac{1}{6}(p^2 - 6m_f^2) + 3 \int_0^1 dx\, M^2 \log\left(\frac{M^2}{\Lambda^2}\right) \right\}
\end{aligned}$$

We will concentrate on the part

$$j(x) := M^2(p^2, x) \log\left(\frac{M^2(p^2, x)}{\Lambda^2}\right).$$

$M^2(p^2, x)$ is defined as:

$$M^2(p^2, x) := p^2 x\bar{x} - m^2 = p^2\left(-x^2 + x + \frac{m_f^2}{p^2}\right).$$

$$\Rightarrow j(x) = p^2\left(-x^2 + x + \frac{m_f^2}{p^2}\right) \log\left(\frac{p^2\left(-x^2 + x + \frac{m_f^2}{p^2}\right)}{\Lambda^2}\right) \tag{B18}$$

$$= p^2\left(-x^2 + x + \frac{m_f^2}{p^2}\right) \left\{ \log\left(-x^2 + x + \frac{m_f^2}{p^2}\right) + \log\left(\frac{p^2}{\Lambda^2}\right) \right\}. \tag{B19}$$

These types of integrals are summarized at the beginning of this appendix.

$$J(p^2, m_\varphi^2, m_G^2, \Lambda) = \frac{1}{4\pi^2}\left\{\Lambda^2 + \frac{1}{6}(p^2 - 6m_f^2) + 3p^2\Delta C_3(c = \frac{m_f^2}{p^2})\right.$$
$$\left. + \frac{3}{6}(p^2 - 6m_f^2)\log\left(\frac{p^2}{\Lambda^2}\right)\right\}$$
$$= \frac{1}{4\pi^2}\left\{\Lambda^2 - \log(\Lambda^2)\left(\frac{1}{2}p^2 - 3m_f^2\right)\right.$$
$$\left. - m_f^2 + p^2\left(\frac{1}{6} + 3\Delta C_3\right) + \left(\frac{1}{2}p^2 - 3m_f^2\right)\log\left(p^2\right)\right\}$$
$$= \frac{1}{4\pi^2}\left\{m^2\left(\frac{\Lambda^2}{m^2} + 3\frac{m_f^2}{m^2}\left(\log\left(\Lambda^2\right) - \log\left(p^2\right)\right) - \frac{m_f^2}{m^2}\right)\right.$$
$$\left. + p^2\left(\frac{1}{2}\log\left(p^2\right) - \frac{1}{2}\log\left(\Lambda^2\right) + 3\Delta C_3 + \frac{1}{6}\right)\right\}$$

$$\Delta C_3(p^2, c = \frac{m_f^2}{p^2}) = \frac{m_f^2}{p^2} - \frac{5}{3} + \frac{11}{12}\log\left(\frac{m_f^2}{p^2}\right)$$
$$+ \frac{1}{8}\left(7 - \frac{4m_f^2}{p^2}\right)\sqrt{1 + \frac{4m_f^2}{p^2}}\log\left(\frac{\sqrt{1 + \frac{4m_f^2}{p^2}} + 1}{\sqrt{1 + \frac{4m_f^2}{p^2}} - 1}\right).$$

Appendix C: Two Particle Energy Levels

This appendix summarized some technical aspects and lists numerical data which is related to the computation of the scattering phases. Table C2 shows the two particle energies obtained in the centre of mass frame. The analysis has been performed with different number of operators in the correlation matrix. Starting from the operator with the lowest energy eigenvalue, the number of operators is increased successively such that the determined error from the Jackknife analysis stayed below 10 %.

A modification of the original work of Lüscher [46] was performed in [54]. The analysis in the centre of mass frame necessitates rather large lattice volumes which is numerically very demanding. The modified method incorporates a moving frame and allows to compute relevant scattering phases already at moderate lattice volumes such as 24^3. An overview of the energy eigenvalues obtained from the moving frame is listed in table C3.

Table C2: The table below lists the obtained set of two particle energies in the centre of mass frame. The number of observables are chosen such that the determined error is below 10 %. The three different physical situations are identified by the three different values of the bare quartic couping.

	$\lambda = 0.01$	$\lambda = 1.0$	$\lambda = \infty$
$L_s = 12$	0.155(3) 0.295(5)	0.233(9) 0.420(8)	0.229(8) 0.45(1) 1.960(1)
$L_s = 16$	0.161(3) 0.280(5)	0.24(1) 0.393(9)	0.242(4) 0.44(2) 1.538(1)
$L_s = 18$	0.164(3) 0.277(4)	0.255(7) 0.399(9)	0.250(8) 0.43(1) 1.4(5)
$L_s = 20$	0.162(4) 0.29(1)	0.27(2) 0.395(8) 0.62(2)	0.265(6) 0.42(1) 1.20(3)
$L_s = 24$	0.166(4) 0.27(1)	0.24(2) 0.380(9) 0.59(2)	0.27(1) 0.43(2) 0.64(3) 0.96(7)
$L_s = 32$	0.171(8) 0.285(8)	0.25(1) 0.40(1) 0.47(1) 0.58(2) 0.76(3)	0.25(1) 0.40(1)
$L_s = 36$	0.167(9) 0.28(3) 0.36(1)	0.27(1) 0.38(1) 0.44(3) 0.70(3)	
$L_s = 40$	0.162(7) 0.283(8)	0.28(2) 0.37(2) 0.42(2) 0.50(3) 0.63(2)	0.23(2) 0.43(3) 0.44(4) 0.47(3) 0.58(5)

Table C3: The table below list the obtained set of two particle energies in the moving frame. Energy values which exeed the elastic region are grey as they cannot be used to compute the scattering phase. The three different physical situations are identified by the three different values of the bare quartic couping in the lattice notation.

	$\lambda = 0.01$	$\lambda = 1.0$	$\lambda = \infty$
$L_s = 12$	0.570(8)	0.62(2) 0.72(1)	0.617(5) 0.753(7)
$L_s = 16$	0.492(4)	0.511(5) 0.585(5)	0.507(4) 0.607(8)
$L_s = 18$	0.42(1) 0.45(1)	0.470(6) 0.539(4)	0.48(2) 0.557(6)
$L_s = 20$	0.396(1) 0.423(4)	0.459(6) 0.525(8) 0.81(3)	0.449(7) 0.538(6)
$L_s = 24$	0.347(5) 0.383(5)	0.418(6) 0.49(1) 0.70(2) 0.79(1)	0.42(1) 0.498(9) 0.69(2)
$L_s = 32$	0.296(7) 0.343(6) 0.64(1)		0.442(7)
$L_s = 36$	0.270(5) 0.329(4)	0.347(6) 0.413(5)	
$L_s = 40$	0.256(3) 0.328(5)	0.337(9) 0.413(8)	0.42(2)

Acknowledgments

The reminiscence of the last years are manifold and include some of the most evocative moments in my life. The following lines and words are devoted to those unique individuals to whom I feel indebted.

Beginning with my first day, Dr. Karl Jansen accompanied me with his profound wisdom on the topic and supported me throughout this work. Apart of his direct and brilliant supervision, he patiently conceded the time and the freedom to deepen my understanding in quantum field theory. For all the hours he spend with me sharing his thoughts, I thank him sincerely.

I equally thank Prof. Michael Müller-Preußker. Working within his group offered a fruitful and pleasant environment. Prof. Michael Müller-Preußker provided unique opportunities to meet and listen to inspiring talks of renowned personalities in physics. Furthermore, he enabled me to visit the summer school in Benasque and to take part in a conference about Higgs physics in Lisbon. Over the last years, he was a wise and supportive guide.

The work on this thesis offered the opportunity to meet some unique and precious companions without whom, this time would have been lonesome. I owe Marina Marinkovic my special thanks. She is a colleague, a physicist and most of all, a near friend. I also thank my former office mate and friend, Sebastian Klein, with whom I ever since share my thoughts and doubts.

I also want to thank some members of the working group whose exceptional personality inspired me: Xu Feng, Jenifer Gonzalez Lopez and Marc Wagner.

Furthermore, I sincerely thank my friend, Björn Andres. His advice and support has been present throughout the years and the discussions with him have deepened my thoughts. I also thank my friends Patrick Gonzalez, Daniel Koschade and Karl Waninger. Being in their company is invaluable.

I especially want to thank Philipp Gerhold. During uncountable hours of discussions with him, I gathered precious knowledge, not only in physics. His exceptional way of thinking defines him and encouraged me to revise what I believed to have understood. Furthermore, I am thankful that I could profit from his genuine work on the simulation algorithm.

John Bulava recently joined this project and his experience in elementary particle physics and curious questions about my work were a great support. I thank him cordially for the time he spend reading my thesis and providing valuable comments and hints.

The numerical computations were performed at "Norddeutscher Verbund für Hoch- und Höchstleistungsrechnen" and I am thankful for the good and friendly support as well as the computing resources granted for this project. Furthermore, I thank "Deutsche Forschungsgemeinschaft" which supported my work and gave the opportunity to present the results at their annual conference. In the last three years, I used the wonderful technical and social infrastructure of the Humboldt Universität zu Berlin and DESY in Zeuthen. Both institutes offered a perfect environment for discussions with other PhD students and experienced physicists.

Finally, I want to thank my beloved friend, partner and wife, Martina Schad. Her patience and guidance gave me the sufficient time and motivation to complete this work. With her brilliant perception, she has reviewed my thoughts and encouraged all my efforts.

These last lines are devoted to my parents, Aleyama Kallarackal and Jos Kallarackal. I thank them for their constant support and love during all periods in my life.

Bibliography

[1] T. Aaltonen et al. Search for New Bottomlike Quark Pair Decays $Q\overline{Q} \to (tW^\mp)(\bar{t}W^\pm)$ in Same-Charge Dilepton Events. *Phys. Rev. Lett.*, 104:091801, 2010. arXiv:0912.1057 [hep-ex].

[2] P. A. Aarnio et al. Measurement of the Mass and Width of the Z^0 Particle from Multi - Hadronic Final States Produced in e^+e^- Annihilations. *Phys. Lett.*, B231:539, 1989.

[3] G. S. Abrams et al. Initial Measurements of Z Boson Resonance Parameters in e+ e- Annihilation. *Phys. Rev. Lett.*, 63:724, 1989.

[4] G. S. Abrams et al. Measurements of Z Boson Resonance Parameters in e^+e^- Annihilation. *Phys. Rev. Lett.*, 63:2173, 1989.

[5] B. Adeva et al. A Determination of the Properties of the Neutral Intermediate Vector Boson Z^0. *Phys. Lett.*, B231:509, 1989.

[6] M. Aizenman. Proof of the Triviality of phi d4 Field Theory and Some Mean-Field Features of Ising Models for d¿4. *Phys. Rev. Lett.*, 47:886–886, 1981.

[7] M. Z. Akrawy et al. Measurement of the Z^0 Mass and Width with the OPAL Detector at LEP. *Phys. Lett.*, B231:530, 1989.

[8] C. Amsler et al. Review of particle physics. *Phys. Lett.*, B667:1, 2008.

[9] R. Barate et al. Search for the standard model Higgs boson at LEP. *Phys. Lett.*, B565:61–75, 2003. arXiv:0306033 [hep-ex].

[10] R. M. Barnett et al. Review of particle physics. Particle Data Group. *Phys. Rev.*, D54:1–720, 1996.

[11] B. Blossier, M. Della Morte, G. von Hippel, T. Mendes, and R. Sommer. On the generalized eigenvalue method for energies and matrix elements in lattice field theory. *JHEP*, 04:094, 2009. arXiv:0902.1265 [hep-lat].

[12] L.S. Brown. *Quantum field theory, Chapter 6*. Cambridge Univ Pr, 1994.

[13] CDF-D0. *Combined CDF and D0 Upper Limits on Standard Model Higgs- Boson Production with up to 6.7 fb^{-1} of Data*, 2010. arXiv:1007.4587 [hep-ex].

[14] M. S. Chanowitz. Higgs Mass Constraints on a Fourth Family: Upper and Lower Limits on CKM Mixing. *Phys. Rev.*, D:035018, 2010. arXiv:1007.0043 [hep-ph].

[15] M. S. Chanowitz, M. A. Furman, and I. Hinchliffe. Weak Interactions of Ultraheavy Fermions. 2. *Nucl. Phys.*, B153:402, 1979.

[16] A. G. Cohen, D. B. Kaplan, and A. E. Nelson. Progress in electroweak baryogenesis. *Ann. Rev. Nucl. Part. Sci.*, 43:27–70, 1993. arXiv:9302210 [hep-ph].

[17] S. R. Coleman and E. J. Weinberg. Radiative Corrections as the Origin of Spontaneous Symmetry Breaking. *Phys. Rev.*, D7:1888–1910, 1973.

[18] D. Decamp et al. Determination of the Number of Light Neutrino Species. *Phys. Lett.*, B231:519, 1989.

[19] Xu Feng, Karl Jansen, and Dru B. Renner. Resonance Parameters of the rho-Meson from Lattice QCD. *Phys. Rev.*, D83:094505, 2011. arXiv:1011.5288 [hep-lat].

[20] Z. Fodor, K. Holland, J. Kuti, D. Nogradi, and C. Schroeder. New Higgs physics from the lattice. *PoS*, LAT2007:056, 2007. arXiv:0710.3151 [hep-lat].

[21] P. H. Frampton, P. Q. Hung, and M. Sher. Quarks and leptons beyond the third generation. *Phys. Rept.*, 330:263, 2000. arXiv:9903387 [hep-ph].

[22] R. Frezzotti and K. Jansen. A polynomial hybrid Monte Carlo algorithm. *Phys. Lett.*, B402:328–334, 1997. arXiv:9702016 [hep-lat].

[23] R. Frezzotti and K. Jansen. The PHMC algorithm for simulations of dynamical fermions. I: Description and properties. *Nucl. Phys.*, B555:395–431, 1999. arXiv:9808011 [hep-lat].

[24] R. Frezzotti and K. Jansen. The PHMC algorithm for simulations of dynamical fermions. II: Performance analysis. *Nucl. Phys.*, B555:432–453, 1999. arXiv:9808038 [hep-lat].

[25] P. Gerhold. Upper and lower Higgs boson mass bounds from a chirally invariant lattice Higgs-Yukawa model. *Ph.D. Thesis*, 2010. arXiv:1002.2569 [hep-lat].

[26] P. Gerhold and K. Jansen. On the phase structure of a chiral invariant Higgs-Yukawa model. *PoS*, LAT2006:043, 2006. arXiv:0610012 [hep-lat].

[27] P. Gerhold and K. Jansen. The phase structure of a chirally invariant lattice Higgs-Yukawa model. *PoS*, LAT2007:075, 2007. arXiv:0710.1106 [hep-lat].

[28] P. Gerhold and K. Jansen. The phase structure of a chirally invariant lattice Higgs-Yukawa model - numerical simulations. *JHEP*, 10:001, 2007. arXiv:0707.3849 [hep-lat].

[29] P. Gerhold and K. Jansen. The phase structure of a chirally invariant lattice Higgs-Yukawa model for small and for large values of the Yukawa coupling constant. *JHEP*, 09:041, 2007. arXiv:0705.2539 [hep-lat].

[30] P. Gerhold and K. Jansen. Lower Higgs boson mass bounds from a chirally invariant lattice Higgs-Yukawa model with overlap fermions. *JHEP*, 07:025, 2009. arXiv:0902.4135 [hep-lat].

[31] P. Gerhold and K. Jansen. Upper Higgs boson mass bounds from a chirally invariant lattice Higgs-Yukawa model. *JHEP*, 04:094, 2010. arXiv:1002.4336 [hep-lat].

[32] P. Gerhold, K. Jansen, and J. Kallarackal. Effects of a potential fourth fermion generation on the upper and lower Higgs boson mass bounds. *PoS*, LAT2010:051, 2010. arXiv:1010.6005 [hep-lat].

[33] P. Gerhold, K. Jansen, and J. Kallarackal. Higgs boson mass bounds in the presence of a very heavy fourth quark generation. *JHEP*, 01:143, 2011. arXiv:1011.1648 [hep-lat].

[34] P. H. Ginsparg and K. G. Wilson. A Remnant of Chiral Symmetry on the Lattice. *Phys. Rev.*, D25:2649, 1982.

[35] Göckeler, M. and Kastrup, H. A. and Westphalen, J. and Zimmermann, F. Scattering phases on finite lattices in the broken phase of the four-dimensional O(4) phi**4 theory. *Nucl. Phys.*, B425:413–448, 1994. arXiv:9402011 [hep-lat].

[36] A. Hasenfratz et al. Finite Size Effects And Spontaneously Broken Symmetries:The Case Of The 0(4) Model. *Z. Phys.*, C46:257, 1990.

[37] A. Hasenfratz et al. Goldstone bosons and finite size effects: A Numerical study of the O(4) model. *Nucl. Phys.*, B356:332–366, 1991.

[38] W. Heisenberg. *Der Teil und das Ganze:*. Harper torchbooks. The Academy library. Harper & Row, 1971.

[39] Hernandez, P. and Jansen, K. and Lüscher, M. Locality properties of Neuberger's lattice Dirac operator. *Nucl. Phys.*, B552:363–378, 1999. arXiv:9808010 [hep-lat].

[40] B. Holdom. The discovery of the fourth family at the LHC: What if? *JHEP*, 08:076, 2006. arXiv:0606146 [hep-ph].

[41] W. Hou. CP Violation and Baryogenesis from New Heavy Quarks. *Chin. J. Phys.*, 47:134, 2009. arXiv:0803.1234 [hep-ph].

[42] C. Itzykson and J.B. Zuber. *Quantum field theory*. Dover books on physics. Dover Publications, 2006.

[43] B. W. Lee, C. Quigg, and H. B. Thacker. Weak Interactions at Very High-Energies: The Role of the Higgs Boson Mass. *Phys. Rev.*, D16:1519, 1977.

[44] LEP-EWWG. *Precision electroweak measurements and constraints on the Standard Model*, 2010. FERMILAB-TM-2480-PPD.

[45] Lüscher, M. Volume Dependence of the Energy Spectrum in Massive Quantum Field Theories. 2. Scattering States. *Commun. Math. Phys.*, 105:153–188, 1986.

[46] Lüscher, M. Two particle states on a torus and their relation to the scattering matrix. *Nucl. Phys.*, B354:531–578, 1991.

[47] Lüscher, M. Exact chiral symmetry on the lattice and the Ginsparg- Wilson relation. *Phys. Lett.*, B428:342–345, 1998. arXiv:9802011 [hep-lat].

[48] Lüscher, M. and Weisz, P. Scaling Laws and Triviality Bounds in the Lattice phi**4 Theory. 3. N Component Model. *Nucl. Phys.*, B318:705, 1989.

[49] Lüscher, M. and Wolff, U. How To Calculate The Elastic Scattering Matrix In Two- Dimensional Quantum Field Theories By Numerical Simulation. *Nucl. Phys.*, B339:222–252, 1990.

[50] I. Montvay and G. Münster. *Quantum fields on a lattice.* Cambridge monographs on mathematical physics. Cambridge University Press, 1994.

[51] H. Neuberger. More about exactly massless quarks on the lattice. *Phys. Lett.*, B427:353–355, 1998. arXiv:9801031 [hep-lat].

[52] W. Nolting. *Grundkurs Theoretische Physik. 5. Quantenmechanik.-Teil 2. Methoden und Anwendungen.* Springer, 1993.

[53] M.E. Peskin and D.V. Schroeder. *An introduction to quantum field theory.* Advanced book program. Addison-Wesley Pub. Co., 1995.

[54] K. Rummukainen and Steven A. Gottlieb. Resonance scattering phase shifts on a nonrest frame lattice. *Nucl. Phys.*, B450:397–436, 1995. arXiv:9503028 [hep-lat].

[55] A. D. Sakharov. Violation of CP Invariance, c Asymmetry, and Baryon Asymmetry of the Universe. *Pisma Zh. Eksp. Teor. Fiz.*, 5:32–35, 1967.

[56] S. Weinberg. *The Quantum Theory of Fields: Supersymmetry.* The Quantum Theory of Fields. Cambridge University Press, 2000.

[57] U. Wolff. Monte Carlo errors with less errors. *Comput. Phys. Commun.*, 156:143–153, 2004. arXiv:0306017 [hep-lat].

Own Publications

[1] Gerhold, P. and Jansen, K. and Kallarackal, J. Higgs Boson Mass Bounds from a Chirally Invariant Lattice Higgs-Yukawa Model.
Springer High Performance Computing in Science and Engineering'10 (2011) Part 1:85-102.

[2] Gerhold, P. and Jansen, K. and Kallarackal, J. Higgs boson mass bounds in the presence of a very heavy fourth quark generation.
JHEP (2011) 1101:143. arXiv:1011.1648 [hep-lat].

[3] Gerhold, P. and Jansen, K. and Kallarackal, J. "Effects of a potential fourth fermion generation on the upper and lower Higgs boson mass bounds",
PoS LATTICE2010 (2010) 051. arXiv:1010.6005 [hep-lat].

[4] Gerhold, P. and Jansen, K. and Kallarackal, J. "Higgs mass bounds from a chirally invariant lattice Higgs-Yukawa model with overlap fermions",
PoS LATTICE2008 (2008) 067. arXiv:0810.4447 [hep-lat].

[5] Giffels, M. and Kallarackal, J. and Kramer, M. and O'Leary, B. and Stahl, A. The lepton-flavour violating decay $\tau \to \mu\mu\bar{\mu}$ at the LHC.
Phys. Rev. D77 (2008). arXiv:0802.0049v2 [hep-ph].

List of Figures

1.1	Particle content of the standard model.	3
2.1	Qualitative phase diagram	28
2.2	Sketch of contour integral for the time correlator	33
2.3	The fermion correlator and effective masses	37
3.1	Feynman diagrams contributing to the scalar self energy	45
3.2	Scalar propagators from continuum Euclidean perturbation theory.	51
3.3	Goldstone propagators from continuum Euclidean perturbation theory.	55
3.4	Comparison: Lattice perturbation theory at $lambda = 0$	59
3.5	Comparison: Lattice perturbation theory at $lambda = 0$	60
4.1	Experimental exclusion plot.	62
4.2	The established results for the lower Higgs boson mass bound.	66
4.3	The established results for the upper Higgs boson mass bound.	67
4.4	Infinite volume extrapolation.	80
4.5	Infinite volume extrapolation of GEVP Higgs mass.	84
4.6	Scattering phases.	87
4.7	Resonance width versus the renoralized quartic coupling	88
5.1	Infinite volume extrapolation at $m_{t'} = 700$ GeV	102
5.2	Physical heavy quark masses at varying cut off values.	103
5.3	Physical Higgs mass bounds in SM4.	104
5.4	Finite size effects at constant cut off of 1500 GeV.	106
5.5	Final result on the Higgs mass bounds in SM4.	107

6.1 Total cross section of the Higgs boson. 114
6.2 Physical Higgs mass bounds in SM4. 115

List of Tables

3.1	Wick contractions contributing to the Higgs boson self energy	43
3.2	Wick contractions contributing to the Goldstone boson self energy	52
4.1	Simulation parameters for the scattering phases.	79
4.2	Lattice volumes for resonance analysis	81
4.3	Comparison of the resonance mass with different approaches.	85
5.1	Bare simulation parameters for the cut off dependency of M_H.	101
5.2	Simulation parameters at $\lambda_0 = \infty$ and 200 GeV $\leq m_{t'} \leq$ 700 GeV.	105
6.1	Concluding summary on the determination of the resonance mass.	112
C2	Two particle energies in the centre of mass frame.	134
C3	Two particle energies in the moving frame.	135

i want morebooks!

Buy your books fast and straightforward online - at one of world's fastest growing online book stores! Environmentally sound due to Print-on-Demand technologies.

Buy your books online at
www.get-morebooks.com

Kaufen Sie Ihre Bücher schnell und unkompliziert online – auf einer der am schnellsten wachsenden Buchhandelsplattformen weltweit! Dank Print-On-Demand umwelt- und ressourcenschonend produziert.

Bücher schneller online kaufen
www.morebooks.de

VDM Verlagsservicegesellschaft mbH
Heinrich-Böcking-Str. 6-8
D - 66121 Saarbrücken

Telefon: +49 681 3720 174
Telefax: +49 681 3720 1749

info@vdm-vsg.de
www.vdm-vsg.de

Printed by Books on Demand GmbH, Norderstedt / Germany